Herbert L. Rice

The theory and practice of interpolation

Including mechanical quadrature and other important problems concerned with the tabular values of functions

Herbert L. Rice

The theory and practice of interpolation
Including mechanical quadrature and other important problems concerned with the tabular values of functions

ISBN/EAN: 9783743335394

Manufactured in Europe, USA, Canada, Australia, Japa

Cover: Foto ©berggeist007 / pixelio.de

Manufactured and distributed by brebook publishing software
(www.brebook.com)

Herbert L. Rice

The theory and practice of interpolation

THE
THEORY AND PRACTICE
OF
INTERPOLATION:

INCLUDING

MECHANICAL QUADRATURE, AND OTHER IMPORTANT PROBLEMS CONCERNED WITH THE TABULAR VALUES OF FUNCTIONS.

WITH THE REQUISITE TABLES.

BY

HERBERT L. RICE, M. S.,

ASSISTANT IN THE OFFICE OF THE AMERICAN EPHEMERIS,
AND PROFESSOR OF ASTRONOMY IN THE CORCORAN SCIENTIFIC SCHOOL, WASHINGTON, D.C.

LYNN, MASS.
THE NICHOLS PRESS — THOS. P. NICHOLS.
1899.

PREFACE.

In preparing the following treatise the author has attempted no marked originality, either of subject matter or method. Indeed, sufficient has hitherto been written of Interpolation, Quadratures, etc., to firmly dissuade one from such an endeavor. Yet of the numerous contributions to these allied subjects, there has appeared thus far no distinct treatise covering the entire ground. As a consequence the author has repeatedly felt the need of a work which would give — exclusive of other matter — a simple, practical, yet comprehensive discussion of all that is useful concerning Differences, Interpolation, Tabular Differentiation and Mechanical Quadrature; — a work, moreover, which would include all tables appertaining to the text which are required by a practical computer. To supply the want thus conceived, the author offers the present volume.

But while viewing the matter in this practical sense, the writer regards his work as no mere compilation. Many of the processes and developments are original, so far as he is concerned, and possibly altogether new; while the same remark applies to a few of the minor *results*. In fact, if adverse criticism be forthcoming, it will probably result largely from the somewhat unusual or individual methods which in many instances have been employed in preference to the customary forms of analysis. On the other hand the author realizes fully the extent of his indebtedness to previous writers for valuable ideas and suggestions; and he desires especially to mention the works of Boole, Chauvenet, Encke, Loomis, Newcomb, and Sawitsch as most valuable sources of information, to which frequent reference has been made.

Concerning the bibliographical list at the close of this volume (which includes the foregoing names), it is but proper to state that references to several of the earliest writers — such as Briggs, Wallis, Mouton, Cotes, Stirling, Mayer, Walmesley, Lalande — have purposely been omitted because of the general inaccessibility of their works. As regards the writings of the present century, however, the author believes that all contributions of importance have been included, and trusts that any omissions of consequence hereafter detected will be regarded merely as oversights.

Special care has been given to the preparation and printing of the tables, with the hope of securing absolute accuracy. At a considerable cost of labor, and by wholly independent methods, the computations were all made in duplicate; and in every case the tabular values are true to the *nearest* unit of the last place. Though a few of these tables have appeared before, several are here published for the first time, and it is hoped they will prove useful to the computer.

In conclusion, the author desires to express his cordial thanks and appreciation to Mr. E. C. Ruebsam, of the Nautical Almanac Office, and to Mr. M. E. Porter, of the Naval Observatory, for much valuable service and many useful suggestions received during the various phases of preparation of this treatise. Feelings of gratitude further inspire — simple justice even demands — a special word in commendation of the publishers, whose uniform courtesy, accuracy and skill have done much to enhance the general value of the work.

H. L. R.

Washington, D.C., *December*, 1899.

CONTENTS.

CHAPTER I.

OF DIFFERENCES.

CHAPTER II.

OF INTERPOLATION.

CHAPTER III.

DERIVATIVES OF TABULAR FUNCTIONS.

CHAPTER IV.

OF MECHANICAL QUADRATURE.

CHAPTER V.

MISCELLANEOUS PROBLEMS AND APPLICATIONS.

APPENDIX.

ON THE SYMBOLIC METHOD OF DEVELOPMENT.

TABLES.

CHAPTER I.

OF DIFFERENCES.

1. In many applications of the exact sciences, and of Astronomy in particular, it is often necessary to tabulate a series of numerical values of some quantity or function, corresponding to certain assumed values of the element or argument upon which the functional values depend.

In the more purely mathematical tables, the function is analytically known; the argument is then the independent variable of the given expression. The common tables of logarithms, trigonometrical functions, squares, cubes, and reciprocals, are examples of tabular functions of this class.

A second and larger class includes those functions which are not related analytically to the argument, but which are either determined directly by experiment, or based wholly or partly upon observation. The final results are usually obtained from the fundamental observations by suitable mathematical transformations or reductions, which frequently include the process of adjustment known as the method of least-squares. Empirical values are also occasionally introduced in the development of functions of this class, to supply some theoretical deficiency.

In the great majority of such cases, the *time* is the argument of the tabulated function. This is particularly the case in astronomical tables. Thus the *Nautical Almanac* gives the right-ascensions and declinations of the sun and the planets for every Greenwich mean noon; in the case of the moon, these coördinates are given for every hour, because of the rapid motion of our satellite. The moon's horizontal parallax is tabulated for every twelve hours; the sun's for every ten days.

In like manner, the readings of the barometer and thermometer are recorded for certain hours of the day, and therefore may be regarded as functions of the time. The velocity of the wind, the height of tide-water, the correction and rate of a clock, are further instances of a large number of physical quantities which are tabulated as functions of the time.

As examples of tabular functions of the physical or observational kind, whose arguments are elements other than the time, we may mention:

(*a*) The force of gravity (determined by pendulum experiments), as a function of the latitude;

(*b*) The atmospheric pressure (determined by the barometer), as a function of the altitude;

(*c*) The angle of refraction in a particular substance, as a function of the angle of incidence.

Although differing thus fundamentally in the character of their respective functions, all mathematical tables are alike in giving the numerical values of the functions for certain assumed values of the argument, so chosen that intermediate values of the function may readily be derived by the process of *interpolation*. For this purpose it is convenient, though not essential, to have the assumed argument values proceed according to some law; and since as a rule the greatest simplicity is attained where the argument varies uniformly, it is nearly always so taken. The *interval* of the argument is decided in general by the rapidity with which the given function varies.

We shall assume throughout these pages that the given values of the argument are equidistant.

The present chapter will be devoted to the subject of *differences*, as defined below. The student should become thoroughly and practically familiar with this fundamental portion of the work before entering upon the chapters that follow.

2. *Definitions and Notation.*—If we have given a series of quantities proceeding according to any law, and take the difference of every two consecutive terms, we obtain a series of values called the *first order of differences*, or briefly, *first differences*.

If we difference the first differences in the same manner, we form a new series called *second differences*. The process may be continued, if necessary, so long as any differences remain.

We shall apply this process of differencing to the tabular values of functions given for equidistant values of the argument.

Let T designate the argument; ω, its interval; $F(T)$, or simply F, the function; $t, t+\omega, t+2\omega, t+3\omega, \ldots\ldots,$ the given values of T; $F_0, F_1, F_2, F_3, \ldots\ldots,$ the corresponding values of $F(T)$; $\Delta', \Delta'', \Delta''', \Delta^{iv}, \ldots\ldots,$ the successive orders of differences. The arrangement is then shown in the following schedule:

Argument	Function	1st Diff.	2d Diff.	3d Diff.	4th Diff.	5th Diff.	6th Diff.
T	$F(T)$	Δ'	Δ''	Δ'''	Δ^{iv}	Δ^{v}	Δ^{vi}
t	F_0						
		a_0					
$t+\omega$	F_1		b_0				
		a_1		c_0			
$t+2\omega$	F_2		b_1		d_0		
		a_2		c_1		e_0	
$t+3\omega$	F_3		b_2		d_1		f_0
		a_3		c_2		e_1	
$t+4\omega$	F_4		b_3		d_2		
		a_4		c_3			
$t+5\omega$	F_5		b_4				
		a_5					
$t+6\omega$	F_6						

where $a_0 = F_1 - F_0$, $a_1 = F_2 - F_1, \ldots;$ $b_0 = a_1 - a_0$, $b_1 = a_2 - a_1, \ldots;$ $c_0 = b_1 - b_0$, $c_1 = b_2 - b_1, \ldots;$ and so on.

We shall also find it convenient to represent $a_0, a_1, a_2, \ldots.$ by $\Delta_0', \Delta_1', \Delta_2', \ldots\ldots,$ respectively; $b_0, b_1, b_2, \ldots.$ by $\Delta_0'', \Delta_1'', \Delta_2'', \ldots\ldots,$ etc., Thus, generally, $\Delta_s^{(n)}$ denotes the $(s+1)^{th}$ value in the column of nth differences.

As an example, we tabulate and difference several successive values of $F(T) \equiv T^4 - 10T^2 - 20$, thus:

T	$F(T)$	Δ'	Δ''	Δ'''	Δ^{iv}	Δ^{v}
0	− 20					
		− 9				
1	− 29		− 6			
		− 15		+ 36		
2	− 44		+ 30		+24	
		+ 15		+ 60		0
3	− 29		+ 90		+24	
		+105		+ 84		0
4	+ 76		+174		+24	
		+279		+108		
5	+355		+282			
		+561				
6	+916					

The differences are in all cases formed by subtracting (algebraically) downwards, as in the above examples. It will be noted that the even differences (Δ'', Δ^{iv}, ...) always fall on the same lines with the argument and function, while the odd differences (Δ', Δ''', Δ^{v}, . . .) lie between the lines.

3. *Method of Checking the Numerical Accuracy of the Differences.* — If, in the numerical example of the last section, we take the algebraic sum of the six given values of Δ', we find

$$-9 - 15 + 15 + 105 + 279 + 561 = +936$$

Subtracting the first value of $F(T)$ from the last, we have

$$+916 - (-20) = +936$$

which agrees with the first result.

Again, in like manner, we find

$$\Delta_0''' + \Delta_1''' + \Delta_2''' = +36 + 60 + 84 = +180 = +174 - (-6) = \Delta_3'' - \Delta_0''$$

These relations may be expressed generally as follows:

THEOREM I. — *The algebraic sum of any* s *consecutive values of* $\Delta^{(n)}$, *is equal to the last, minus the first, of the* $s+1$ *consecutive* $\Delta^{(n-1)}$ *terms used in forming the* s *values of* $\Delta^{(n)}$.

To prove this proposition, let the differences be as below:

$$\begin{array}{llllllll} \Delta^{(n-1)} : & h_1 \quad h_2 \quad h_3 \, . \, . \, . \, . \, . \, . \, . \, h_{s-1} \quad h_s \quad h_{s+1} \\ \Delta^{(n)} \quad : & \quad k_1 \quad k_2 \quad k_3 \, . \, . \, . \, . \, . \, . \, . \, . \, k_{s-1} \; k_s \end{array}$$

Then, from the definition of differences, we have

$$k_1 = h_2 - h_1, \qquad k_2 = h_3 - h_2, \qquad \ldots\ldots, \qquad k_{s-1} = h_s - h_{s-1}, \qquad k_s = h_{s+1} - h_s$$

Hence, by addition, we find

$$k_1 + k_2 + k_3 + \ldots\ldots + k_{s-1} + k_s = h_{s+1} - h_1$$

which is the algebraic statement of Theorem I. This theorem may obviously be applied as an independent check upon the numerical accuracy of the differencing.

4. Theorem II.—*If the differences of N values of $F(T)$ are taken, $N-n$ values of $\Delta^{(n)}$ are derived; it being assumed that $N > n$.*

For, N functions evidently yield $N-1$ values of Δ', $N-2$ values of Δ'', $N-3$ values of Δ''', etc.; hence N values of $F(T)$ yield $N-n$ values of $\Delta^{(n)}$.

5. *Inversion of a Series of Functions.*—It is sometimes necessary or convenient to invert a given column of functions, thus bringing the last value into the position of the first, the next to the last into the position of the second, etc. For example, let us invert the series given in §2, and observe the effect of this inversion upon the differences. Thus we find:

T	F(T)	Δ′	Δ″	Δ‴	Δ^iv	Δ^v
6	+916					
		−561				
5	+355		+282			
		−279		−108		
4	+ 76		+174		+24	
		−105		− 84		0
3	− 29		+ 90		+24	
		− 15		− 60		0
2	− 44		+ 30		+24	
		+ 15		− 36		
1	− 29		− 6			
		+ 9				
0	− 20					

Comparing this table with the original, we first observe that each column of differences is inverted, like the column of functions itself. Further, having regard to signs, we see that the first and third differences (the *odd* orders) have changed signs throughout; while Δ'' and Δ^{iv} (the *even* orders) remain unaltered in sign.

To prove that such an effect is true generally, we consider the two series below, the second series being an inversion of the first :

$F(T)$	Δ'	Δ''	Δ'''	Δ^{iv}
F_0				
	a_0			
F_1		b_0		
	a_1		c_0	
F_2		b_1		d_0
	a_2		c_1	
F_3		b_2		d_1
	a_3		c_2	
F_4		b_3		
	a_4			
F_5				

$F(T)$	Δ'	Δ''	Δ'''	Δ^{iv}
F_5				
	α_0			
F_4		β_0		
	α_1		γ_0	
F_3		β_1		δ_0
	α_2		γ_1	
F_2		β_2		δ_1
	α_3		γ_2	
F_1		β_3		
	α_4			
F_0				

Comparing the first differences, we find

$$\begin{aligned} \alpha_0 &= F_4 - F_5 = -(F_5 - F_4) = -a_4 \\ \alpha_1 &= F_3 - F_4 = -(F_4 - F_3) = -a_3 \\ \alpha_2 &= F_2 - F_3 = -(F_3 - F_2) = -a_2 \\ &\quad \cdot \quad \cdot \quad \cdot \qquad \cdot \quad \cdot \quad \cdot \end{aligned}$$

Hence, for the second differences, we obtain

$$\begin{aligned} \beta_0 &= \alpha_1 - \alpha_0 = -a_3 - (-a_4) = a_4 - a_3 = b_3 \\ \beta_1 &= \alpha_2 - \alpha_1 = -a_2 - (-a_3) = a_3 - a_2 = b_2 \\ &\quad \cdot \quad \cdot \quad \cdot \qquad \cdot \quad \cdot \quad \cdot \quad \cdot \quad \cdot \quad \cdot \end{aligned}$$

Thus, the inversion of the functions inverts Δ', and changes its signs throughout ; whereas Δ'' is inverted, but does not change in sign, Further, since Δ''' and Δ^{iv} have the same relation to Δ'', that Δ' and Δ'' have to $F(T)$, it is manifest that Δ''' inverts and changes signs, while Δ^{iv} inverts with signs unaltered. Extending this reasoning, we have the following proposition :

THEOREM III.—*Inverting a series of functions inverts each column of differences and changes the signs of the odd orders* (Δ', Δ''', Δ^{v}, ...), *while the signs of the even orders* (Δ'', Δ^{iv}, ...) *remain unchanged.*

In practice it is seldom necessary to re-tabulate the function in the inverted order, since we may readily conceive the inversion to be made, merely allowing for the changes of sign in Δ', Δ''', Δ^{v}, etc.

6. THEOREM IV.— *The* n^{th} *differences of the sums of two series of functions are equal to the sums of the corresponding* n^{th} *differences of the two component series.*

To prove generally, let $F_0, F_1, F_2, \ldots,$ and $f_0, f_1, f_2, \ldots,$ denote the two series of functions; then the sums of the two series will be F_0+f_0, F_1+f_1, F_2+f_2, ... Also, let us designate the first differences of these three series by Δ', δ', and D', respectively; their values are hence as follows:

	Δ'		δ'		D'
F_0	F_1-F_0	f_0	f_1-f_0	F_0+f_0	$(F_1+f_1)-(F_0+f_0)$
F_1	F_2-F_1	f_1	f_2-f_1	F_1+f_1	$(F_2+f_2)-(F_1+f_1)$
F_2	⋮	f_2	⋮	F_2+f_2	⋮
⋮		⋮		⋮	

We therefore have

$$D_0' = (F_1+f_1)-(F_0+f_0) = F_1+f_1-F_0-f_0 = (F_1-F_0)+(f_1-f_0) = \Delta_0'+\delta_0'$$
$$D_1' = (F_2+f_2)-(F_1+f_1) = F_2+f_2-F_1-f_1 = (F_2-F_1)+(f_2-f_1) = \Delta_1'+\delta_1'$$
$$\cdots\cdots\cdots$$

These relations prove the theorem directly for $n=1$; but since the second differences are formed from the first differences in the same manner that the latter are derived from the given functions, the theorem is also true for $n=2$. Similarly with the following differences, each order being the first difference of the order just preceding. Hence the theorem is true generally.

As an example we write:

F	Δ'	Δ''	Δ'''	f	δ'	δ''	δ'''	$F+f$	D'	D''	D'''
− 5	+ 1	+12	+6	+14	+2	+1	−4	+ 9	+ 3	+13	+2
− 4	+13	+18	+6	+16	+3	−3	−3	+ 12	+16	+15	+3
+ 9	+31	+24		+19	0	−6		+ 28	+31	+18	
+40	+55			+19	−6			+ 59	+49		
+95				+13				+108			

It will be observed that the values of D', D'' and D''' are in accord with the theorem.

7. *Irregularities in the Differences.*—In the numerical example of §2, the values of Δ^{v} are all zero. In such a case, we say that the differences are perfectly smooth or *regular*. In practice, however, the

differences frequently exhibit a small degree of irregularity, owing to the omission of decimals in the approximate values of the functions employed. As an example, we take the following values of T^4, true to the nearest unit of the second decimal :

T	$F(T)\equiv T^4$	Δ'	Δ''	Δ'''	Δ^{iv}
2.0	16.00				
		+3.45			
2.1	19.45		+0.53		
		3.98		+0.04	
2.2	23.43		.57		+0.04
		4.55		.08	
2.3	27.98		.65		− .05
		5.20		.03	
2.4	33.18		.68		+ .05
		5.88		.08	
2.5	39.06		.76		− .04
		6.64		.04	
2.6	45.70		.80		+0.05
		7.44		+0.09	
2.7	53.14		+0.89		
		+8.33			
2.8	61.47				

That the irregularity here manifest in the outer differences is due to the fact that the tabular values are only approximate (not the true mathematical values of the function), may easily be shown by Theorem IV, thus : let

$\bar{F}$ denote the true value of the function ;
F, its approximate value as above ;
$f = F - \bar{F}$, the difference of these values.

Then, since F is given to the nearest unit of the second place, f may have any value from -0.5 to $+0.5$, in terms of the same unit. Moreover, the values of f do not follow any law of progression, but proceed at random, with arbitrary changes of sign. Hence, the *differences* of f will be irregular. The differences of $\bar{F}$ must proceed regularly, however, since $\bar{F}$ is the true mathematical value of a continuous function. Now, since $F = \bar{F} + f$, it follows from Theorem IV that the differences of F must equal the sums of the corresponding differences of $\bar{F}$ and f; therefore, *the differences of F must contain just such irregularities as are inevitable in the differences of f.*

To illustrate this principle, we tabulate below the values of $\bar{F}$, along with the given series, F; whence f follows, in units of the second decimal, and also the differences of f to the fourth order :

T	$\bar{F}(T)$	$F(T)$	$f = F - \bar{F}$	Δ'	Δ''	Δ'''	Δ^{iv}
2.0	16.00,00	16.00	0.00				
				+0.19			
2.1	19.44,81	19.45	+0.19		+0.06		
				+0.25		−1.16	
2.2	23.42,56	23.43	+0.44		−1.10		+3.76
				−0.85		+2.60	
2.3	27.98,41	27.98	−0.41		+1.50		−5.24
				+0.65		−2.64	
2.4	33.17,76	33.18	+0.24		−1.14		+4.76
				−0.49		+2.12	
2.5	39.06,25	39.06	−0.25		+0.98		−4.24
				+0.49		−2.12	
2.6	45.69,76	45.70	+0.24		−1.14		+4.76
				−0.65		+2.64	
2.7	53.14,41	53.14	−0.41		+1.50		
				+0.85			
2.8	61.46,56	61.47	+0.44				

We now bring together, from the above tables, the fourth differences of F and f, denoting these quantities by $(\Delta^{iv})F$ and $(\Delta^{iv})f$, respectively. The fourth differences of $\bar{F}$ then follow, since we have shown that $(\Delta^{iv})F = (\Delta^{iv})\bar{F} + (\Delta^{iv})f$; thus we form the table below:

$(\Delta^{iv})F$	$(\Delta^{iv})f$	$(\Delta^{iv})\bar{F}$
+0.04	+0.03,76	+0.0024
−0.05	−0.05,24	+0.0024
+0.05	+0.04,76	+0.0024
−0.04	−0.04,24	+0.0024
+0.05	+0.04,76	+0.0024

It will be observed that the fourth differences of $\bar{F}(T)$ are absolutely uniform,—that is, the irregularities in $(\Delta^{iv})F$ and $(\Delta^{iv})f$ exactly correspond, or balance. The slight irregularity in the outer differences of the series $F(T)$ is therefore due entirely to the omission of decimals, since it wholly disappears when we employ the true mathematical values, $\bar{F}(T)$.

As a valuable exercise, the student should now difference the function $\bar{F}$ directly, and find the fourth differences exactly as above deduced.

8. *Detection of Accidental Errors.*—We have just seen how a slight deviation from the true value of a tabular function will manifest itself by means of irregularities in the differences. If, then, some one value of a series is in error by an appreciable quantity, an inspection of the differences will indicate definitely the location and magnitude of the error sought.

To investigate the principle that underlies the method, let

$$F_0,\ F_1,\ F_2,\ F_3,\ F_4,\ F_5,\ \ldots\ldots$$

denote the *correct* values of any function $F(T)$ (tabulated for equidistant values of T), and let the differences be as shown in the schedule below :

$F(T)$	Δ'	Δ''	Δ'''	Δ^{iv}	Δ^{v}
F_0	a_0	b_0	c_0	d_0	e_0
F_1	a_1	b_1	c_1	d_1	e_1
F_2	a_2	b_2	c_2	d_2	e_2
F_3	a_3	b_3	c_3	d_3	e_3
F_4	a_4	b_4	c_4	d_4	e_4
F_5	a_5	b_5	c_5	d_5	e_5
F_6	a_6	b_6	c_6	d_6	e_6
F_7	a_7	b_7	c_7	d_7	e_7
F_8	a_8	b_8	c_8	d_8	
F_9	a_9	b_9	c_9		
F_{10}	a_{10}	b_{10}			
F_{11}	a_{11}				
F_{12}					

Let us now assume that some one function, say F_6, is in error by the quantity ϵ, so that $F_6+\epsilon$ is tabulated in place of the true value F_6; the differences of the incorrect series will therefore be found as follows :

$F(T)+\epsilon$	Δ'	Δ''	Δ'''	Δ^{iv}	Δ^{v}
F_0	a_0	b_0	c_0	d_0	e_0
F_1	a_1	b_1	c_1	d_1	$e_1+\epsilon$
F_2	a_2	b_2	c_2	$d_2+\epsilon$	$e_2-5\epsilon$
F_3	a_3	b_3	$c_3+\epsilon$	$d_3-4\epsilon$	$e_3+10\epsilon$
F_4	a_4	$b_4+\epsilon$	$c_4-3\epsilon$	$d_4+6\epsilon$	$e_4-10\epsilon$
F_5	$a_5+\epsilon$	$b_5-2\epsilon$	$c_5+3\epsilon$	$d_5-4\epsilon$	$e_5+5\epsilon$
$F_6+\epsilon$	$a_6-\epsilon$	$b_6+\epsilon$	$c_6-\epsilon$	$d_6+\epsilon$	$e_6-\epsilon$
F_7	a_7	b_7	c_7	d_7	e_7
F_8	a_8	b_8	c_8	d_8	
F_9	a_9	b_9	c_9		
F_{10}	a_{10}	b_{10}			
F_{11}	a_{11}				
F_{12}					

Now, because the differences of the correct table contain no irregularities, we see that the differences of the incorrect table consist of series of regular values, to which are alternately added and subtracted the terms in ϵ, shown in the above schedule. The law of progression and increase in the coefficients of ϵ, along the successive orders of differences, is easily seen to be that of the binomial coefficients, with alternate signs. Hence, in practice, we have only to carry the differencing to that order at which the differences of the correct functions would vanish, or sensibly so; the location and magnitude of the error will then be clearly shown by a succession of $+$ and $-$ terms, following the binomial law.

Thus, if the values of Δ^{v} vanish in the correct table above, the fifth differences of the incorrect series will be 0, $+\epsilon$, -5ϵ, $+10\epsilon$, -10ϵ, $+5\epsilon$, $-\epsilon$, 0; the initial value, $+\epsilon$, is therefore the error sought, both as to magnitude and sign. The required function is found by tracing backwards and downwards along the line of heavy type from $e_1+\epsilon$ to $F_6+\epsilon$, which is the incorrect function; and since the *correction* is the negative of the error, we have $(F_6+\epsilon)-\epsilon$, or F_6, for the true value of the function in question.

9. We shall now consider several examples, in order that the process may be fully understood.

EXAMPLE I.—Find the error in the following table of $F(T)\equiv T^3$:

T	$F(T)\equiv T^3$	Δ'	Δ''	Δ'''	Δ^{iv}	c	$\Delta^{iv}+c$
1	1						
		+7					
2	8		+12				
		19		+6			
3	27		18		0		0
		37		+6			
4	64		24		**−10**	+10	0
		61		**−4**			
5	125		**20**		+40	−40	0
		81		+36			
6	**206**		56		−60	+60	0
		137		−24			
7	343		**32**		+40	−40	0
		169		**+16**			
8	512		48		**−10**	+10	0
		217		+6			
9	729		+54				
		+271					
10	1000						

The differencing is continued until we find a complete alternation of signs, as in Δ^{iv}. Now the binomial coefficients of the fourth order are 1, 4, 6, 4, 1; it is also seen that the values of Δ^{iv} are just these numbers multiplied by 10. Hence, an error of 10 units exists somewhere in the function F; its location is easily determined by tracing backwards and downwards along the line of -10, -4, $+20$, $+81$, to the number 206, which is the quantity sought. The required function is also found by tracing backwards and upwards along the line of -10, $+16$, $+32$, $+137$, to 206; in practice, both lines should be followed, to guard against mistake.

Finally, the number 206 is *too small* by 10 units, since the *sign* of the error is shown by the leading or initial value of the binomial series in Δ^{iv}, namely, —10. A correction of +10 is therefore to be applied to the incorrect function, giving 216 for its true value.

In the column c, following Δ^{iv} in the above table, are given the corrections to Δ^{iv}, due to the correction of +10 to the function. The column $\Delta^{iv}+c$ therefore gives the 4th differences of the true or corrected series. It is always well to re-difference the series after a correction has been applied, to check the accuracy of the work.

EXAMPLE II.—Find the error in the following table of logarithms:

T	$\log T$	Δ'	Δ''	Δ'''	c	$\Delta'''+c$
45	1.6532					
		+458				
50	1.6990		−44			
		414		**+13**	− 5	+8
55	1.7404		**31**			
		383		−10	+15	5
. 60	**1.7787**		41			
		342		+21	−15	6
65	1.8129		**20**			
		322		− **2**	+ 5	3
70	1.8451		22			
		300		+ 2		+2
75	1.8751		−20			
		+280				
80	1.9031					

The third differences are here sufficient to point out the error; the correction given under c appears to improve Δ''' in the best manner, thus indicating that log 60 should be 1.7782 instead of 1.7787. It will be observed that a correction of —6 is nearly as efficient as —5 in the above case, and that —5.5 is better than either; this is because the value of log 60 to five places is 1.77815.

EXAMPLE III.—Correct the error in the following ephemeris of the moon's latitude:

Date 1898	Moon's Lat. β	Δ'	Δ''	Δ'''	Δ^{iv}	Δ^{v}	c	$\Delta^{v}+c$
	° ′ ″	′ ″	′ ″	″	″	″	″	″
May 8.5	−1 59 54.2							
		+37 10.0						
9.0	1 22 44.2		+1 7.2					
		38 17.2		−42.7				
9.5	0 44 27.0		+0 24.5		+ 1.7			
		38 41.7		41.0		+ **14.1**	− 12.8	+1.3
10.0	−0 5 45.3		−0 16.5		**+15.8**			
		38 25.2		**25.2**		− 63.4	+ 64.0	0.6
10.5	+0 32 39.9		**0 41.7**		−47.6			
		37 43.5		−72.8		+128.2	−128.0	0.2
11.0	**1 10 23.4**		1 54.5		+80.6			
		35 49.0		+ 7.8		−127.5	+128.0	+0.5
11.5	1 46 12.4		**1 46.7**		−46.9			
		34 2.3		**−39.1**		. . . . .	. . .	. . .
12.0	2 20 14.7		−2 25.8		. . . .			
		+31 36.5				. . . . .	. . .	. . .
12.5	+2 51 51.2							

In this example the error is readily indicated in Δ^{v}, for which order the binomial coefficients are 1, 5, 10, 10, 5, 1. Although but four values of Δ^{v} are available, these are here sufficient. A slight inspection shows that a correction of —13″.0, as applied to the latitude for May 11.0, will very nearly serve the purpose; —13″.0 being a trifle too great numerically, we soon find by trial that —12″.8 produces the best result. Hence, the moon's latitude for May 11.0 should read, +1° 10′ 10″.6.

10. *Correction of Errors when More than One Function is Affected.* — Thus far we have considered examples of an error in one function only. When two or more consecutive or neighboring values are in error, the problem of correction becomes more complicated and difficult. It may even become indeterminate in some cases, since only *accidental* errors can be detected by the differences. Several successive functions, and possibly all, may contain systematic errors which do not affect the regularity of the differences.

In general, the correction of a group of errors by differences may be considered practicable only when the law of the function is not obscured or altered by the presence of those errors. More definitely, the method may be regarded as available in the case of two or perhaps three neighboring functions, provided the errors be accidental in character, and of sufficient magnitude to produce a distinct and definitive irregularity in the differences.

EXAMPLE I. — Correct the errors in the following tabulation of $F(T) \equiv 2T^3 - 25T - 40$:

T	$F(T)$	Δ'	Δ''	Δ'''	c_1	$\Delta'''+c_1$	c_2	$\Delta'''+c_1+c_2$
−4	− 68							
		+ 49						
3	19		−36					
		+ 13		+12		+12		+12
2	6		24					
		− 11		12		12		12
−1	17		−12					
		23		12		12		12
0	40		0					
		23		7	+ 5	12		12
+1	63		+ 7					
		− 16		27	−15	12		12
2	79		34					
		+ 18		+ 5	+15	+20	− 8	12
3	61		39					
		57		− 7	− 5	−12	+24	12
4	− 4		32					
		89		+36		+36	−24	12
5	+ 85		68					
		157		4		4	+ 8	12
6	242		72					
		229		+12		+12		+12
7	471		+84					
		+313						
+8	+784							

We carry the differences to the third order, and note that the first three values of Δ''' are constant, and equal to +12; hence, in column c_1, we place the correction of +5. This gives a corrected series for Δ''', shown under $\Delta'''+c_1$. The latter column clearly indicates a correction of —8, as applied in c_2; this gives a final corrected column of third differences, with the constant value of +12. Hence, the value $F(T)$ for $T=+2$, should read —74 instead of —79; for $T=+4$, we should have —12 instead of —4.

Example II. — Correct the errors which occur in the following ephemeris of the sun's declination:

Date 1898	Sun's Decl. δ	Δ'	Δ''	Δ'''	c_1 & c_2	$\Delta'''+c_1+c_2$	c_3	$\Delta'''+c_1+c_2+c_3$
	° ′ ″	′ ″	″	″		″		″
Jan. 28	−18 6 34.7							
		+32 30.7						
30	17 34 4.0		+74.3					
		33 45.0		− 3.2		− 3.2		−3.2
Feb. 1	17 0 19.0		71.1					
		34 56.1		− 3.1		− 3.1		3.1
3	16 25 22.9		68.0					
		36 4.1		+ 0.1	−3.2	− 3.1		3.1
5	15 49 18.8		68.1					
		37 12.2		− 7.7	+9.6	+ 1.9	− 5.1	3.2
7	15 12 6.6		60.4					
		38 12.6		−11.8	−9.6+3.0	−18.4	+15.3	3.1
9	14 33 54.0		48.6					
		39 1.2		+18.0	+3.2−9.0	+12.2	−15.3	3.1
11	13 54 52.8		66.6					
		40 7.8		−17.5	+9.0	− 8.5	+ 5.1	3.4
13	13 14 45.0		49.1					
		40 56.9		− 0.3	−3.0	− 3.3		3.3
15	12 33 48.1		48.8					
		41 45.7		− 3.5		− 3.5		−3.5
17	11 52 2.4		+45.3					
		+42 31.0						
19	−11 9 31.4							

In this case, the first, second, and last values of Δ''' are —3.2, —3.1 and —3.5, respectively, thus indicating a decided uniform tendency in the third differences. The first function in error is clearly the value for Feb. 7, and the last, that for Feb. 11. There may be an uncertainty of a unit or two in the values of their corrections at the outset; a few trials, however, will indicate that —3.2 is the best value to apply to +0.1 in Δ''', and +3.0 to the term —11.8. By means of these corrections, the first three and the last two values of Δ''' are brought into practical uniformity. In the column c_1 & c_2 are given the corrections to Δ''', according to the binomial numbers, 1, 3, 3, 1. In the next column, the sum $\Delta'''+c_1+c_2$ is written, which evidently requires a third correction, tabulated under c_3.

The differences are now sufficiently smooth. Since c_3 corresponds to a correction of —5″.1 to δ for Feb. 9, we conclude that the correct values of δ for Feb. 7, 9, and 11, should read, —15° 12′ 9″.8, —14° 33′ 59″.1, and —13° 54′ 49″.8, respectively.

It occasionally happens that some order of difference clearly indicates a correction corresponding to the binomial coefficients of a lower order than that of the difference in question. This means the existence of an error in some earlier order of *difference*, rather than an error in the column of functions. For example, if Δ^{v} requires a correction of the order 1, 3, 3, 1, it follows that an error exists in Δ'', since Δ^{v} is the *third* difference of Δ''. More generally, when $\Delta^{(n)}$ requires a correction according to the binomial coefficients of the mth order, an error exists in $\Delta^{(n-m)}$. These remarks imply the necessity of some caution on the part of the beginner.

It will be observed that when either the first or last function of a series is in error, only the first or the last term in each order of difference will be affected, and only by an amount numerically equal to the error. Hence, in such cases, the method above explained is of little value.

In general, it may be stated that when errors have been discovered by differencing, it is advisable to re-compute the values in question, when the data for the calculation are available.

General Properties of Differences.

11. Let $F(t)$, $F(t+\omega)$, $F(t+2\omega)$, ... represent any series of tabular functions, whose differences are taken as in the schedule below:

Function, $F(T)$	Δ'	Δ''	Δ'''	. . .	. . .	$\Delta^{(n)}$	$\Delta^{(n+1)}$	. . .
$F(t)$								
	Δ_0'							
$F(t+\omega)$		Δ_0''						
	Δ_1'		Δ_0'''					
$F(t+2\omega)$		Δ_1''		. . .				
	Δ_2'		Δ_1'''		. . .			
$F(t+3\omega)$	.	Δ_2''		. . .		$\Delta_0^{(n)}$		
.	.	.	Δ_2'''		. . .		$\Delta_0^{(n+1)}$	
.	.	.	.	. . .		$\Delta_1^{(n)}$		. . .
.	.	.	.	.	. . .		$\Delta_1^{(n+1)}$	
.	.	.	.	.	.	$\Delta_2^{(n)}$		. . .
.	.	.	.	.	.		$\Delta_2^{(n+1)}$	
.	.	.	.	.		.	.	. . .
$F[t+s\omega]$	.	.	.	.		.	.	.
	Δ_s'	.	.	.		.	.	.
$F[t+(s+1)\omega]$		Δ_s''	.	.		.	.	.
	Δ'_{s+1}		Δ_s'''	.	.	.	.	.
$F[t+(s+2)\omega]$	. .	Δ''_{s+1}		. . .	.	.	.	.
.		.	Δ'''_{s+1}		. . .	.	.	.
. .		.		. . .		$\Delta_s^{(n)}$	.	.
.		.		.	. . .		$\Delta_s^{(n+1)}$	.
.		.	.	.		$\Delta_{s+1}^{(n)}$		. . .
.	.	.	.		.	.	$\Delta_{s+1}^{(n+1)}$	
.	.	.	.	.	.	.		. . .

We shall assume that $F(T)$ is a finite and continuous function, and that $F(t+s\omega)$ is capable of expansion in a series of powers of $s\omega$, within the limits of the given table; then, denoting the successive derivatives of $F(T)$ by $F'(T)$, $F''(T)$, etc., we have, by TAYLOR'S Theorem, the following expressions :

$$\left.\begin{array}{lcl}
F(t) & = & F(t) \\
F(t+\omega) & = & F(t) + \omega F'(t) + \dfrac{\omega^2}{\lfloor 2} F''(t) + \dfrac{\omega^3}{\lfloor 3} F'''(t) + \dfrac{\omega^4}{\lfloor 4} F^{\text{iv}}(t) + \ldots \\
F(t+2\omega) & = & F(t) + 2\omega F'(t) + 4\dfrac{\omega^2}{\lfloor 2} F''(t) + 8\dfrac{\omega^3}{\lfloor 3} F'''(t) + 16\dfrac{\omega^4}{\lfloor 4} F^{\text{iv}}(t) + \ldots \\
F(t+3\omega) & = & F(t) + 3\omega F'(t) + 9\dfrac{\omega^2}{\lfloor 2} F''(t) + 27\dfrac{\omega^3}{\lfloor 3} F'''(t) + 81\dfrac{\omega^4}{\lfloor 4} F^{\text{iv}}(t) + \ldots \\
F(t+4\omega) & = & F(t) + 4\omega F'(t) + 16\dfrac{\omega^2}{\lfloor 2} F''(t) + 64\dfrac{\omega^3}{\lfloor 3} F'''(t) + 256\dfrac{\omega^4}{\lfloor 4} F^{\text{iv}}(t) + \ldots \\
\ldots & & \ldots
\end{array}\right\} \quad (0)$$

Differencing these values of the functions in the usual manner, we obtain successively the expressions for Δ', Δ'', Δ''' . . . ., as follows :

$$\left.\begin{array}{lcl}
\Delta_0' & = & \omega F'(t) + \dfrac{\omega^2}{\lfloor 2} F''(t) + \dfrac{\omega^3}{\lfloor 3} F'''(t) + \dfrac{\omega^4}{\lfloor 4} F^{\text{iv}}(t) + \ldots \\
\Delta_1' & = & \omega F'(t) + 3\dfrac{\omega^2}{\lfloor 2} F''(t) + 7\dfrac{\omega^3}{\lfloor 3} F'''(t) + 15\dfrac{\omega^4}{\lfloor 4} F^{\text{iv}}(t) + \ldots \\
\Delta_2' & = & \omega F'(t) + 5\dfrac{\omega^2}{\lfloor 2} F''(t) + 19\dfrac{\omega^3}{\lfloor 3} F'''(t) + 65\dfrac{\omega^4}{\lfloor 4} F^{\text{iv}}(t) + \ldots \\
\Delta_3' & = & \omega F'(t) + 7\dfrac{\omega^2}{\lfloor 2} F''(t) + 37\dfrac{\omega^3}{\lfloor 3} F'''(t) + 175\dfrac{\omega^4}{\lfloor 4} F^{\text{iv}}(t) + \ldots \\
\ldots & & \ldots
\end{array}\right\} \quad (1)$$

$$\left.\begin{array}{lcl}
\Delta_0'' & = & \omega^2 F''(t) + \omega^3 F'''(t) + \frac{7}{12}\omega^4 F^{\text{iv}}(t) + \ldots \\
\Delta_1'' & = & \omega^2 F''(t) + 2\omega^3 F'''(t) + \frac{25}{12}\omega^4 F^{\text{iv}}(t) + \ldots \\
\Delta_2'' & = & \omega^2 F''(t) + 3\omega^3 F'''(t) + \frac{55}{12}\omega^4 F^{\text{iv}}(t) + \ldots \\
\ldots & & \ldots
\end{array}\right\} \quad (2)$$

$$\left.\begin{array}{lcl}
\Delta_0''' & = & \omega^3 F'''(t) + \frac{3}{2}\omega^4 F^{\text{iv}}(t) + \ldots \\
\Delta_1''' & = & \omega^3 F'''(t) + \frac{5}{2}\omega^4 F^{\text{iv}}(t) + \ldots \\
\ldots & & \ldots
\end{array}\right\} \quad (3)$$

It will be observed that all terms of the expansions (0) are of the general form, $K\omega^r F^{(r)}(t)$; where K denotes a numerical factor, and r an integer which increases by unity as we proceed from any term to the next term following. Hence, the *differences* will contain only terms of this form. We thus see, *a priori*, that any difference of the nth order must be of the form

$$\Delta_t^{(n)} = A\omega^r F^{(r)}(t) + B\omega^{r+1} F^{(r+1)}(t) + C\omega^{r+2} F^{(r+2)}(t) + D\omega^{r+3} F^{(r+3)}(t) + \ldots .$$

Let us now *assume* what appears from (1), (2), and (3) to be the general law; that is

$$A = 1 \qquad r = n$$

leaving the coefficients $B, C, D, \ldots.$ undetermined for the present. We therefore assume

$$\Delta_t^{(n)} = \omega^n F^{(n)}(t) + B\omega^{n+1} F^{(n+1)}(t) + C\omega^{n+2} F^{(n+2)}(t) + D\omega^{n+3} F^{(n+3)}(t) + \ldots . \qquad (4)$$

Since the value of t is arbitrary, we may write $t+\omega$ for t; by making this substitution in the right-hand member of (4), we evidently get the expression for the nth difference immediately following $\Delta_t^{(n)}$, — that is, the value of $\Delta_{t+1}^{(n)}$. Hence we have

$$\Delta_{t+1}^{(n)} = \omega^n F^{(n)}(t+\omega) + B\omega^{n+1} F^{(n+1)}(t+\omega) + C\omega^{n+2} F^{(n+2)}(t+\omega) + D\omega^{n+3} F^{(n+3)}(t+\omega) + \ldots$$

Developing the functions of the right-hand member by TAYLOR'S Theorem, we find

$$\begin{aligned}\Delta_{t+1}^{(n)} = \omega^n &\left[F^{(n)}(t) + \omega F^{(n+1)}(t) + \frac{\omega^2}{\underline{|2}} F^{(n+2)}(t) + \frac{\omega^3}{\underline{|3}} F^{(n+3)}(t) + \ldots\right] \\ &+ B\omega^{n+1}\left[F^{(n+1)}(t) + \omega F^{(n+2)}(t) + \frac{\omega^2}{\underline{|2}} F^{(n+3)}(t) + \ldots\right] \\ &+ C\omega^{n+2}\left[F^{(n+2)}(t) + \omega F^{(n+3)}(t) + \ldots\right] \\ &+ D\omega^{n+3}\left[F^{(n+3)}(t) + \ldots\right] \qquad + \ldots\end{aligned}$$

Collecting the coefficients of $F^{(n)}(t)$, $F^{(n+1)}(t)$, ..., we obtain

$$\begin{aligned}\Delta_{t+1}^{(n)} = \omega^n F^{(n)}(t) + (B+1)\,\omega^{n+1} F^{(n+1)}(t) + \left(C + B + \frac{1}{\underline{|2}}\right)\omega^{n+2} F^{(n+2)}(t) \qquad (5) \\ + \left(D + C + \frac{B}{\underline{|2}} + \frac{1}{\underline{|3}}\right)\omega^{n+3} F^{(n+3)}(t) + \ldots\end{aligned}$$

Subtracting (4) from (5), and observing that $\Delta_{t+1}^{(n)} - \Delta_t^{(n)} = \Delta_t^{(n+1)}$, we get

$$\begin{aligned}\Delta_t^{(n+1)} = \omega^{n+1} F^{(n+1)}(t) + \left(B + \frac{1}{\underline{|2}}\right)\omega^{n+2} F^{(n+2)}(t) + \left(C + \frac{B}{\underline{|2}} + \frac{1}{\underline{|3}}\right)\omega^{n+3} F^{(n+3)}(t) \\ + \left(D + \frac{C}{\underline{|2}} + \frac{B}{\underline{|3}} + \frac{1}{\underline{|4}}\right)\omega^{n+4} F^{(n+4)}(t) + \ldots\end{aligned}$$

If, therefore, we put

$$\left.\begin{aligned} B' &= B+\frac{1}{\underline{\vert 2}} \\ C' &= C+\frac{B}{\underline{\vert 2}}+\frac{1}{\underline{\vert 3}} \\ D' &= D+\frac{C}{\underline{\vert 2}}+\frac{B}{\underline{\vert 3}}+\frac{1}{\underline{\vert 4}} \\ & \dots\dots\dots\dots \end{aligned}\right\} \qquad (6)$$

we have

$$\Delta_s^{(n+1)} = \omega^{n+1}F^{(n+1)}(t)+B'\omega^{n+2}F^{(n+2)}(t)+C'\omega^{n+3}F^{(n+3)}(t)+D'\omega^{n+4}F^{(n+4)}(t)+\dots \qquad (7)$$

Hence, if the general form of expression assumed in (4) is true for the index n, it follows from (7) that it is also true for $n+1$; but we see by equations (1), (2), and (3), that the law obtains for $n = 1, 2, 3$, respectively; hence it holds for $n = 4$; and so on indefinitely. The expression (4) is therefore true for all positive integral values of n.

12. We have now to determine the coefficients $B, C, D, \dots,$ of equation (4). These quantities are evidently functions of n and s, and will be determined in the following manner:

First, we take $s = 0$, and determine the constants for $\Delta_0^{(n)}$, which we shall denote for this purpose by $B_n, C_n, D_n, \dots$

These values are found by induction, thus: the relations (6) give $B_{n+1}, C_{n+1}, D_{n+1}, \dots$ in terms of $B_n, C_n, D_n, \dots$ Making $n = 1$, we take $B_1, C_1, D_1, \dots$ directly from the first of the equations (1); a continued application of (6) therefore gives successively the values of $B_2, B_3, B_4, \dots B_{n-1}, B_n$. Similarly, we derive $C_n, D_n, \dots$ Hence, the coefficients of (4) become known for $s = 0$.

Second, the coefficients of $\Delta_s^{(n)}$ easily follow from those of $\Delta_0^{(n)}$; for it is clear from the schedule of §11 that $\Delta_s^{(n)}$ is related to $F(t+s\omega)$ in precisely the manner that $\Delta_0^{(n)}$ is related to $F(t)$. Hence, if for brevity we write

$$\Delta_0^{(n)} = \Psi(t)$$

we shall have, since the value of t is arbitrary,

$$\Delta_s^{(n)} = \Psi(t+s\omega)$$

Then, expanding $\Psi(t+s\omega)$ in a series of powers of $s\omega$, we arrive at an expression of the form (4), in which the coefficients are fully determined functions of n and s.

To perform the steps indicated, we take from the first of the equations (1) the following values:

$$B_1 = \tfrac{1}{2} \qquad C_1 = \tfrac{1}{6} \qquad D_1 = \tfrac{1}{24} \tag{8}$$

To find B_n: By repeated application of the first of (6), we have

$$\begin{aligned} B_2 &= B_1 + \tfrac{1}{2} \\ B_3 &= B_2 + \tfrac{1}{2} \\ B_4 &= B_3 + \tfrac{1}{2} \\ &\cdots\cdots \\ B_{n-1} &= B_{n-2} + \tfrac{1}{2} \\ B_n &= B_{n-1} + \tfrac{1}{2} \end{aligned}$$

Hence, by the addition of these $n-1$ equations, we get

$$B_n = B_1 + \tfrac{1}{2}(n-1) = \tfrac{1}{2} + \tfrac{1}{2}(n-1) = \frac{n}{2} \tag{9}$$

To find C_n: Using the second of (6), we obtain

$$\begin{aligned} C_2 &= C_1 + \tfrac{1}{2} B_1 + \tfrac{1}{6} \\ C_3 &= C_2 + \tfrac{1}{2} B_2 + \tfrac{1}{6} \\ &\cdots\cdots\cdots \\ C_n &= C_{n-1} + \tfrac{1}{2} B_{n-1} + \tfrac{1}{6} \end{aligned}$$

whence, by addition, we find

$$C_n = C_1 + \tfrac{1}{2}(B_1 + B_2 + \ldots + B_{n-1}) + \tfrac{1}{6}(n-1)$$

Since $C_1 = \tfrac{1}{6}$, this gives

$$C_n = \tfrac{1}{2}(B_1 + B_2 + \ldots + B_{n-1}) + \frac{n}{6} = \tfrac{1}{2}\sum_{r=1}^{r=n-1} B_r + \frac{n}{6}$$

But, from (9), we have $B_r = \frac{r}{2}$; hence we get

$$C_n = \tfrac{1}{4}\sum_{r=1}^{r=n-1} r + \frac{n}{6} = \tfrac{1}{4}\left[\frac{n(n-1)}{2}\right] + \frac{n}{6} = \frac{n}{24}(3n+1) \tag{10}$$

To find D_n: Again, from (6), we derive

$$\begin{aligned} D_2 &= D_1 + \tfrac{1}{2} C_1 + \tfrac{1}{6} B_1 + \tfrac{1}{24} \\ D_3 &= D_2 + \tfrac{1}{2} C_2 + \tfrac{1}{6} B_2 + \tfrac{1}{24} \\ &\cdots\cdots\cdots\cdots \\ D_n &= D_{n-1} + \tfrac{1}{2} C_{n-1} + \tfrac{1}{6} B_{n-1} + \tfrac{1}{24} \end{aligned}$$

whence

$$D_n = D_1 + \tfrac{1}{2}\sum_{r=1}^{r=n-1} C_r + \tfrac{1}{6}\sum_{r=1}^{r=n-1} B_r + \tfrac{1}{24}(n-1) = \tfrac{1}{2}\sum_{r=1}^{r=n-1} C_r + \frac{n^2}{24}$$

From (10), we have

$$C_r = \frac{r}{24}(3r+1) = \frac{r^2}{8} + \frac{r}{24}$$

$$\therefore D_n = \tfrac{1}{16}\sum_{r=1}^{r=n-1} r^2 + \tfrac{1}{48}\sum_{r=1}^{r=n-1} r + \frac{n^2}{24} = \tfrac{1}{16}\left[\frac{n}{6}(n-1)(2n-1)\right] + \tfrac{1}{48}\left[\frac{n(n-1)}{2}\right] + \frac{n^2}{24}$$

or

$$D_n = \frac{n^2}{48}(n+1) \tag{11}$$

In like manner, the process might be extended to the values of E_n, F_n, ... ; but the results already obtained are here sufficient. Substituting in equation (4) the values of B_n, C_n, and D_n, given by (9), (10), (11), (remembering that these values suppose $s=0$), we have

$$\Delta_0^{(n)} = \omega^n F^{(n)}(t) + \frac{n}{2}\,\omega^{n+1}F^{(n+1)}(t) + \frac{n}{24}(3n+1)\,\omega^{n+2}F^{(n+2)}(t) + \frac{n^2}{48}(n+1)\,\omega^{n+3}F^{(n+3)}(t) + \ . \ . \tag{12}$$

We now obtain from (12) the expression for $\Delta_s^{(n)}$. As already proposed, we write

$$\Delta_0^{(n)} = \Psi(t) = \omega^n F^{(n)}(t) + B_n\,\omega^{n+1}F^{(n+1)}(t) + C_n\,\omega^{n+2}F^{(n+2)}(t) + \ . \ . \ . \ .$$

Then, as shown above, we shall have

$$\begin{aligned}\Delta_s^{(n)} &= \Psi(t+s\omega) = \Psi(t) + s\omega\,\Psi'(t) + \frac{s^2\omega^2}{\lfloor 2}\Psi''(t) + \frac{s^3\omega^3}{\lfloor 3}\Psi'''(t) + \ . \ . \ . \ . \\ &= \left(\omega^n F^{(n)}(t) + B_n\omega^{n+1}F^{(n+1)}(t) + C_n\omega^{n+2}F^{(n+2)}(t) + D_n\omega^{n+3}F^{(n+3)}(t) + \ . \ . \ . \ .\right) \\ &\quad + s\omega\left(\omega^n F^{(n+1)}(t) + B_n\omega^{n+1}F^{(n+2)}(t) + C_n\omega^{n+2}F^{(n+3)}(t) + \ . \ . \ . \ .\right) \\ &\quad + \frac{s^2\omega^2}{\lfloor 2}\left(\omega^n F^{(n+2)}(t) + B_n\omega^{n+1}F^{(n+3)}(t) + \ . \ . \ . \ .\right) \\ &\quad + \frac{s^3\omega^3}{\lfloor 3}\left(\omega^n F^{(n+3)}(t) + \ . \ . \ . \ .\right) \ + \ . \ . \ . \ .\end{aligned}$$

Upon arranging this expression according to ascending powers of ω, we get

$$\begin{aligned}\Delta_s^{(n)} = \omega^n F^{(n)}(t) + (B_n + s)\,\omega^{n+1}F^{(n+1)}(t) + \left(C_n + B_n s + \frac{s^2}{\lfloor 2}\right)\omega^{n+2}F^{(n+2)}(t) \\ + \left(D_n + C_n s + \frac{B_n s^2}{\lfloor 2} + \frac{s^3}{\lfloor 3}\right)\omega^{n+3}F^{(n+3)}(t) + \ . \ . \ . \ .\end{aligned} \tag{13}$$

Hence, substituting the foregoing values of B_n, C_n, and D_n, we find that the values of $B, C, D, \ldots.$ in equation (4) are as follows:

$$\left.\begin{aligned} B &= \frac{n}{2} + s \\ C &= \frac{n}{24}(3n+1) + \frac{s}{2}(n+s) \\ D &= \left(\frac{n+2s}{12}\right)\left[\frac{n(n+1)}{4} + s(n+s)\right] \\ &\ldots\ldots\ldots\ldots \end{aligned}\right\} \qquad (14)$$

These results are easily verified by substituting special values of n and s, and comparing with the coefficients in equations (1), (2), (3); thus, putting $s = 1$, and taking $n = 1, 2, 3$, successively, we obtain the numerical coefficients in the expansions of Δ_1', Δ''_1, and Δ_1''', respectively.

13. *Remarkable Formal Relation between the Expressions for $\Delta_0^{(n)}$ and Δ_0'.*—The coefficients $B_n, C_n, D_n, \ldots.$, in the expression for $\Delta_0^{(n)}$, may also be determined by the following method, which not only is shorter than the above, but also possesses the advantage of showing a direct relation between the expressions for $\Delta_0^{(n)}$ and Δ_0', respectively. Retaining the above notation, we write (12) in the form

$$\Delta_0^{(n)} = \omega^n F^{(n)}(t) + B_n \omega^{n+1} F^{(n+1)}(t) + C_n \omega^{n+2} F^{(n+2)}(t) + \ldots. \qquad (15)$$

We now let

$$\phi_n(y) \equiv y^n + B_n y^{n+1} + C_n y^{n+2} + D_n y^{n+3} + \ldots. \qquad (15a)$$

be an auxiliary expression, such that the coefficient of y^{n+r} is the coefficient of $\omega^{n+r} F^{(n+r)}(t)$ in (15). Writing $n+1$ for n in (15a), and using the relations (6), we have

$$\begin{aligned} \phi_{n+1}(y) = y^{n+1} + \left(B_n + \frac{1}{\underline{|2}}\right) y^{n+2} + \left(C_n + \frac{B_n}{\underline{|2}} + \frac{1}{\underline{|3}}\right) y^{n+3} \qquad (16) \\ + \left(D_n + \frac{C_n}{\underline{|2}} + \frac{B_n}{\underline{|3}} + \frac{1}{\underline{|4}}\right) y^{n+4} + \ldots. \end{aligned}$$

Again, since the coefficients of $\phi_1(y)$ are those of Δ_0', we obtain from (1),

$$\phi_1(y) = y + \frac{y^2}{\underline{|2}} + \frac{y^3}{\underline{|3}} + \frac{y^4}{\underline{|4}} + \ldots. \qquad (17)$$

By re-arranging the terms of (16), we find

$$
\begin{aligned}
q_{n+1}(y) = \; & y^{n+1} + \frac{y^{n+2}}{\underline{|2}} + \frac{y^{n+3}}{\underline{|3}} + \frac{y^{n+4}}{\underline{|4}} + \ldots . \\
& + B_n\left(y^{n+2} + \frac{y^{n+3}}{\underline{|2}} + \frac{y^{n+4}}{\underline{|3}} + \frac{y^{n+5}}{\underline{|4}} + \ldots .\right) \\
& + C_n\left(y^{n+3} + \frac{y^{n+4}}{\underline{|2}} + \frac{y^{n+5}}{\underline{|3}} + \frac{y^{n+6}}{\underline{|4}} + \ldots .\right) \\
& + D_n\left(y^{n+4} + \frac{y^{n+5}}{\underline{|2}} + \frac{y^{n+6}}{\underline{|3}} + \frac{y^{n+7}}{\underline{|4}} + \ldots .\right) \\
& + \ldots \ldots \ldots \ldots \\
= \; & y^n\left(y + \frac{y^2}{\underline{|2}} + \frac{y^3}{\underline{|3}} + \frac{y^4}{\underline{|4}} + \ldots .\right) \\
& + B_n y^{n+1}\left(y + \frac{y^2}{\underline{|2}} + \frac{y^3}{\underline{|3}} + \frac{y^4}{\underline{|4}} + \ldots .\right) \\
& + C_n y^{n+2}\left(y + \frac{y^2}{\underline{|2}} + \frac{y^3}{\underline{|3}} + \frac{y^4}{\underline{|4}} + \ldots .\right) \\
& + D_n y^{n+3}\left(y + \frac{y^2}{\underline{|2}} + \frac{y^3}{\underline{|3}} + \frac{y^4}{\underline{|4}} + \ldots .\right) \\
& + \ldots \ldots \ldots \ldots \\
= \; & \left(y + \frac{y^2}{\underline{|2}} + \frac{y^2}{\underline{|3}} + \frac{y^2}{\underline{|4}} + \ldots .\right)\left(y^n + B_n y^{n+1} + C_n y^{n+2} + D_n y^{n+3} + \ldots .\right)
\end{aligned}
$$

Hence, by (15a) and (17), we have

$$q_{n+1} = q_1 \cdot q_n$$

Taking $n = 1, 2, 3, \ldots . \; n-1$, successively, we find

$$
\begin{array}{llll}
q_2 = q_1 q_1 & & q_6 & = q_1 q_5 \\
q_3 = q_1 q_2 & & \ldots & \\
q_4 = q_1 q_3 & & q_{n-1} & = q_1 q_{n-2} \\
q_5 = q_1 q_4 & & q_n & = q_1 q_{n-1}
\end{array}
$$

Multiplying these equations together member for member, and cancelling the common factors, we obtain

$$q_n = (q_1)^n \tag{18}$$

Therefore, by (17), we have

$$q_n(y) = \left(y + \frac{y^2}{\underline{|2}} + \frac{y^3}{\underline{|3}} + \frac{y^4}{\underline{|4}} + \ldots .\right)^n = y^n\left(1 + \frac{y}{\underline{|2}} + \frac{y^2}{\underline{|3}} + \frac{y^3}{\underline{|4}} + \ldots .\right)^n$$

$$\therefore q_n(y) = y^n + \frac{n}{2}\, y^{n+1} + \frac{n}{24}\,(3n+1)\, y^{n+2} + \frac{n^2}{48}\,(n+1)\, y^{n+3} + \ldots . \tag{19}$$

Comparing coefficients in (15a) and (19), we find

$$B_n = \frac{n}{2} \quad , \quad C_n = \frac{n}{24}(3n+1) \quad , \quad D_n = \frac{n^2}{48}(n+1) , \quad . \ . \ . \ . \qquad (20)$$

Substituting these values in (15), the latter becomes

$$\Delta_0^{(n)} = \omega^n F^{(n)}(t) + \frac{n}{2}\,\omega^{n+1} F^{(n+1)}(t) + \frac{n}{24}(3n+1)\,\omega^{n+2} F^{(n+2)}(t) + \frac{n^2}{48}(n+1)\,\omega^{n+3} F^{(n+3)}(t) + \ . \ .$$

which agrees with (12).

These results may be conveniently expressed symbolically: thus, let us represent the quantities $\Delta_0', \Delta_0'', \Delta_0''', \ . \ . \ . \Delta_0^{(n)}$ by $\Delta_0, \Delta_0^2, \Delta_0^3, \ . \ . \ . \Delta_0^n$; and for $\omega F'(t)$, $\omega^2 F''(t)$, $\omega^3 F'''(t)$, ... $\omega^n F^{(n)}(t)$ let us write the symbols $D, D^2, D^3, \ . \ . \ . \ . \ D^n$, respectively; then we shall have

$$\left.\begin{aligned}
\Delta_0 &= D + \frac{D^2}{\underline{|2}} + \frac{D^3}{\underline{|3}} + \frac{D^4}{\underline{|4}} + \frac{D^5}{\underline{|5}} + \ . \ . \ . \ . \\
\Delta_0^2 &= \left(D + \frac{D^2}{\underline{|2}} + \frac{D^3}{\underline{|3}} + \frac{D^4}{\underline{|4}} + \ . \ . \ . \ . \right)^2 = D^2 + D^3 + \frac{7}{12}D^4 + \frac{1}{4}D^5 + \ . \ . \ . \ . \\
\Delta_0^3 &= \left(D + \frac{D^2}{\underline{|2}} + \frac{D^3}{\underline{|3}} + \frac{D^4}{\underline{|4}} + \ . \ . \ . \ . \right)^3 = D^3 + \frac{3}{2}D^4 + \frac{5}{4}D^5 + \frac{3}{4}D^6 + \ . \ . \ . \ . \\
&. \ . \\
\Delta_0^n &= \left(D + \frac{D^2}{\underline{|2}} + \frac{D^3}{\underline{|3}} + \frac{D^4}{\underline{|4}} + \ . \ . \ . \ . \right)^n \\
&= D^n + \frac{n}{2}D^{n+1} + \frac{n}{24}(3n+1)D^{n+2} + \frac{n^2}{48}(n+1)D^{n+3} + \ . \ . \ . \ .
\end{aligned}\right\} \qquad (21)$$

14. THEOREM V.—*The nth differences of any rational integral expression of the nth degree are constant. If the general form of the function is* $F(T) \equiv \alpha T^n + \beta T^{n-1} + \gamma T^{n-2} + \ . \ . \ . \ .$, *the constant value of* $\Delta^{(n)}$ *is* $\omega^n \alpha \underline{|n}$.

For, from the nature of the function, we have, evidently,

$$F^{(n)}(t) = \frac{d^{(n)}F}{dT^n} = \alpha \underline{|n}$$

and

$$F^{(n+1)}(t) = F^{(n+2)}(t) = \ . \ . \ . \ . \ = 0$$

Hence, from (4), we have

$$\Delta_t^{(n)} = \omega^n F^{(n)}(t) = \omega^n \alpha \underline{|n} \qquad (22)$$

The theorem is therefore true, whatever the value of the constant interval ω. Several examples have already occurred: in §2 we have the differences of $F(T) \equiv T^4 - 10T^2 - 20$; here $n = 4$, $a = 1$, $\omega = 1$. Hence, by (22), we get

$$\Delta^{iv} = \underline{|4} = 24$$

—the value already found by differencing.

In Example I of §9, $F(T) \equiv T^3$, $\omega = 1$; we there obtained for the value of the third difference

$$\Delta''' = 6$$

which agrees with the theorem.

Again, in Example I of §10, $F(T) \equiv 2T^3 - 25T - 40$, $\omega = 1$; whence the theorem requires

$$\Delta''' = a\underline{|3} = 2\underline{|3} = 12$$

which is the result already obtained.

15. THEOREM VI.—*If the nth differences of a series of quantities (tabulated for equidistant values of T) are constant, the given quantities are the tabular values of a rational integral function of the form* $F(T) \equiv aT^n + \beta T^{n-1} + \gamma T^{n-2} + \ldots$.

This proposition is the converse of THEOREM V, and is proved as follows:

Let $F(T)$ denote the function whose *true mathematical values*, tabulated for the given values of T, form the given series of quantities. From (4) and (5), we see that the expressions for $\Delta^{(n)}_{i}$ and $\Delta^{(n)}_{i+1}$ agree only in their first term, $\omega^n F^{(n)}(t)$; the remaining terms of like order in ω having unlike coefficients. Hence, the conditions necessary in order that $\Delta^{(n)}$ shall be constant throughout are as follows:

First, that $\omega^n F^{(n)}(t)$ does *not* vanish;
Second, that $\omega^{n+1} F^{(n+1)}(t) = \omega^{n+2} F^{(n+2)}(t) = \ldots = 0$;

But, since ω cannot vanish, these conditions reduce to the form—

$$\left.\begin{array}{l} F^{(n)}(t) \gtrless 0 \\ F^{(n+1)}(t) = F^{(n+2)}(t) = \ldots = 0 \end{array}\right\} \qquad (23)$$

If now we put

$$T = t + \tau \qquad (24)$$

then, by TAYLOR'S Theorem, we have

$$F(T) = F(t+\tau) = F(t) + \tau F'(t) + \frac{\tau^2}{\underline{|2}} F''(t) + \ldots + \frac{\tau^n}{\underline{|n}} F^{(n)}(t) + \frac{\tau^{n+1}}{\underline{|n+1}} F^{(n+1)}(t) + \ldots$$

By (23), this gives

$$F(T) = F(t) + \tau F'(t) + \ldots + \frac{\tau^{n-1}}{\underline{|n-1}} F^{(n-1)}(t) + \frac{\tau^n}{\underline{|n}} F^{(n)}(t) \qquad (25)$$

in which, we observe, the coefficient of τ^n cannot vanish. Substituting in (25) the value of τ given by (24), we obtain

$$F(T) = F(t) + (T-t)F'(t) + (T-t)^2 \frac{F''(t)}{\underline{|2}} + \ldots + (T-t)^n \frac{F^{(n)}(t)}{\underline{|n}}$$

Since t has a fixed value, the right-hand member of this equation is an expression of the nth degree in the variable T, and hence may be written in the form

$$F(T) \equiv \alpha T^n + \beta T^{n-1} + \gamma T^{n-2} + \ldots \qquad (26)$$

which establishes the theorem.

16. *Convergence of the Differences in Practice.*—In the discussion of Theorems V and VI, we were concerned with the true mathematical values of the quantities involved. In practice, however, the absolute or true mathematical values of functions are seldom employed; frequently, as previously noted, a function is tabulated only to a certain degree of approximation, enough decimals being retained to give the desired accuracy. We observe that in such cases there is a tendency of the differences to decrease numerically, and usually to vanish sensibly, as the order of difference progresses. The explanation of this tendency follows readily from equation (4), thus: for any given function, the derivatives $F^{(n)}(t)$, $F^{(n+1)}(t)$, $F^{(n+2)}(t)$, ... have definite values; hence, the value of ω may be chosen sufficiently small to render all the terms in the second member of (4) insensible, *except the first.* When this condition obtains, the value of $\Delta^{(n)}$ is sensibly constant, and equal to $\omega^n F^{(n)}(t)$. The differences of $F(T)$ are thus practically brought to a termination at the nth order, whether the function is algebraic or transcendental.

In many cases the values of the successive derivatives converge rapidly; the chosen value of ω may then be quite large, and yet allow the differences to sensibly vanish at an early order. This is equivalent to the obvious statement that, when a function is to be tabulated so as to difference readily, the interval of the argument must be decided by the manner in which the given function varies.

To exemplify these principles, we take the following table of seven-figure logarithms:

T	Log T	Δ'	Δ''	Δ'''
1.00	0.0000000			
		+43214		
1.01	.0043214		−426	
		42788		+8
1.02	.0086002		418	
		42370		9
1.03	.0128372		409	
		41961		8
1.04	.0170333		401	
		41560		+7
1.05	.0211893		−394	
		+41166		
1.06	0.0253059			

In this case, $\omega = 0.01$, $t = 1.00$, $t + \omega = 1.01$, $t+2\omega = 1.02$, etc. To serve our present purpose, we here transcribe from (1), (2), and (3), the following expressions:

$$\left.\begin{aligned}
\Delta_0' &= \omega F'(t) + \frac{\omega^2}{2} F''(t) + \frac{\omega^3}{6} F'''(t) + \frac{\omega^4}{24} F^{\text{iv}}(t) + \ldots \\
\Delta_0'' &= \omega^2 F''(t) + \omega^3 F'''(t) + \tfrac{7}{12}\, \omega^4 F^{\text{iv}}(t) + \ldots \\
\Delta_0''' &= \omega^3 F'''(t) + \tfrac{3}{2}\, \omega^4 F^{\text{iv}}(t) + \ldots \\
\Delta_0^{\text{iv}} &= \omega^4 F^{\text{iv}}(t) + \ldots
\end{aligned}\right\} \qquad (27)$$

Since $F(T) = \log T$, we have

$$F'(t) = +Mt^{-1}\,,\quad F''(t) = -Mt^{-2}\,,\quad F'''(t) = +2Mt^{-3}\,,\quad F^{\text{iv}}(t) = -6Mt^{-4}\,,\ \ldots$$

where M is the modulus of the common system of logarithms, $= 0.434294$. Hence, with $t = 1$ and $\omega = 0.01$, we find

$$\omega F'(t) = +0.0043429{,}4 \qquad \omega^3 F'''(t) = +0.0000008{,}7$$
$$\omega^2 F''(t) = -0.0000434{,}3 \qquad \omega^4 F^{\text{iv}}(t) = -0.0000000{,}3$$

Substituting these numerical values in (27), we obtain, in units of the 7th decimal,

$$\Delta_0' = +43214 \qquad \Delta_0'' = -426 \qquad \Delta_0''' = +8 \qquad \Delta_0^{\text{iv}} = 0$$

which agree substantially with the results obtained above by direct differencing. The rapid convergence of the differences is thus seen to be due to the small value of the interval ω, which makes the term $\omega^3 F'''(t)$ appreciable, but renders $\omega^4 F^{\text{iv}}(t)$, $\omega^5 F^{\text{v}}(t)$, ... quite insensible; accordingly, Δ''' is the last difference which we need take into account, the remaining differences being practically zero.

We may add that if the values of T in the present table were 100, 101, etc., instead of the given values, the interval ω would become 1 instead of 0.01, and hence $\omega, \omega^2, \omega^3, \omega^4$, ... would not converge as above. We should then, however, have $t = 100$ instead of 1, which would cause the successive *derivatives* to converge rapidly, as is obvious from the general expression

$$F^{(n)}(t) = (-1)^{n-1} M \underline{|n-1} \cdot \frac{1}{t^n}$$

Furthermore, the differences of $F(T)$ contain only terms of the form

$$K\omega^n F^{(n)}(t) = (-1)^{n-1} KM \underline{|n-1} \left(\frac{\omega}{t}\right)^n$$

where K denotes a numerical factor; hence, since the values of ω and t are both increased one hundred-fold by the assumed change, it is evident that the general term $K\omega^n F^{(n)}(t)$ is not altered thereby. The *differences* are therefore unaltered by the proposed change; this conclusion is confirmed by the consideration that the assumed alteration in T would merely change the logarithmic characteristic from 0 to 2, and thus would not affect the resulting differences. These observations illustrate the case of a tabular function whose successive derivatives converge rapidly, whereby a comparatively large argument interval may be used, and yet allow the resulting series of differences to converge as rapidly as may be required.

17. As a second example, we consider the following table of cubes:

T	T^3	Δ'	Δ''	Δ'''
5.16	137.39			
		+4.03		
5.21	141.42		+0.08	
		4.11		0
5.26	145.53		.08	
		4.19		0
5.31	149.72		.08	
		4.27		0
5.36	153.99		.08	
		4.35		0
5.41	158.34		+0.08	
		+4.43		
5.46	162.77			

We have already seen (Theorem V) that when the true mathematical values of T^3 are tabulated, the third differences are constant, the fourth differences being the first order to vanish. In the present table, however, only two decimals have been retained in T^3, whereas the true value involves six places. To this degree of approximation, the third differences are entirely insensible; this follows from Theorem V, which gives for the constant value of Δ'''—

$$\Delta''' = \omega^3 \alpha \, \underline{|3}$$

In this example we have

$$\omega = 0.05 \qquad \alpha = 1$$

and hence

$$\Delta''' = (0.05)^3 \times 6 = 0.00,075$$

which is insensible when only two decimals are concerned. Thus, in the approximations so frequently used in practice, the differences generally terminate (either absolutely or approximately) at some order earlier than would occur if the true mathematical values of the function were employed.

It may be added that the above example affords an illustration of Theorem VI. For, since the second differences are here absolutely constant, it follows from this theorem that the tabular quantities are the true mathematical values (corresponding to the given values of T) of some function of the form

$$F(T) \equiv \alpha T^2 + \beta T + \gamma$$

Thus, in particular, if the student tabulates the function

$$F(T) \equiv 16\,(T^2 - 5.3325\,T + 9.476975)$$

for $T = 5.16,\ 5.21,\ \ldots\ 5.46,$ and retains all decimals involved, he will find his tabular numbers identical with the above series.

18. *To Express* $\omega^n F^{(n)}(t)$ *in Terms of* $\Delta_0^{(n)}, \Delta_0^{(n+1)}, \Delta_0^{(n+2)},$ *etc.*— The problem consists in reversing the series (15), which expresses $\Delta_0^{(n)}$ in terms of $\omega^n F^{(n)}(t),\ \omega^{n+1} F^{(n+1)}(t),\ \ldots$

Let us denote $\omega^r F^{(r)}(t)$ by x_r; then, writing successively, $n,\ n+1,\ n+2,\ \ldots$ for n in (15), we have

$$\left.\begin{aligned}
\Delta_0^{(n)} &= x_n + B_n x_{n+1} + C_n x_{n+2} + D_n x_{n+3} + \quad \ldots \\
\Delta_0^{(n+1)} &= x_{n+1} + B_{n+1} x_{n+2} + C_{n+1} x_{n+3} + D_{n+1} x_{n+4} + \ldots \\
\Delta_0^{(n+2)} &= x_{n+2} + B_{n+2} x_{n+3} + C_{n+2} x_{n+4} + D_{n+2} x_{n+5} + \ldots \\
&\ldots\ldots\ldots\ldots
\end{aligned}\right\} \tag{28}$$

from which we obtain, by transposition,

$$\left.\begin{aligned}
x_n &= \Delta_0^{(n)} - B_n x_{n+1} - C_n x_{n+2} - D_n x_{n+3} - \quad \ldots \\
x_{n+1} &= \Delta_0^{(n+1)} - B_{n+1} x_{n+2} - C_{n+1} x_{n+3} - D_{n+1} x_{n+4} - \ldots \\
x_{n+2} &= \Delta_0^{(n+2)} - B_{n+2} x_{n+3} - C_{n+2} x_{n+4} - D_{n+2} x_{n+5} - \ldots \\
&\ldots\ldots\ldots\ldots
\end{aligned}\right\} \tag{29}$$

The second of the equations (29) gives a value of x_{n+1}, which, substituted in the first equation, gives x_n in terms of $\Delta_0^{(n)}, \Delta_0^{(n+1)}, x_{n+2}, x_{n+3}, \ldots$; substituting in the latter expression the value of x_{n+2} given by the third of (29), we find x_n in terms of $\Delta_0^{(n)}, \Delta_0^{(n+1)}, \Delta_0^{(n+2)}, x_{n+3}, x_{n+4}, \ldots$. Continuing this process of elimination indefinitely, we arrive at an expression of the form

$$x_n \equiv \omega^n F^{(n)}(t) = \Delta_0^{(n)} + b_n \Delta_0^{(n+1)} + c_n \Delta_0^{(n+2)} + d_n \Delta_0^{(n+3)} + \ldots \tag{30}$$

The coefficients $b_n, c_n, d_n, \ldots$ must now be determined. From (15*a*) we obtain the following group of equations:

$$\left.\begin{aligned}
q_n &= y^n + B_n y^{n+1} + C_n y^{n+2} + D_n y^{n+3} + \quad \ldots \\
q_{n+1} &= y^{n+1} + B_{n+1} y^{n+2} + C_{n+1} y^{n+3} + D_{n+1} y^{n+4} + \ldots \\
q_{n+2} &= y^{n+2} + B_{n+2} y^{n+3} + C_{n+2} y^{n+4} + D_{n+2} y^{n+5} + \ldots \\
&\ldots\ldots\ldots\ldots
\end{aligned}\right\} \tag{31}$$

Comparing (28) and (31), we observe that the latter group may be obtained from the former by writing q_r and y^r for $\Delta_0^{(r)}$ and x_r, respectively; the algebraic relations in both groups are otherwise identical. Hence, if from (31) we seek to express y^n in terms of $q_n, q_{n+1}, q_{n+2}, \ldots$, the process of reversion will be identical with that which gives x_n in terms of $\Delta_0^{(n)}, \Delta_0^{(n+1)}, \ldots$; hence we must find

$$y^n = q_n + b_n q_{n+1} + c_n q_{n+2} + d_n q_{n+3} + \ldots \tag{32}$$

the coefficients being those of (30). Therefore, by (18), we have

$$y^n = q_1^n + b_n q_1^{n+1} + c_n q_1^{n+2} + d_n q_1^{n+3} + \ldots \tag{33}$$

Taking $n = 1$, in (30) and (33), we obtain

$$x_1 = \Delta_0' + b_1 \Delta_0'' + c_1 \Delta_0''' + d_1 \Delta_0^{\text{iv}} + \ldots \tag{34}$$

and

$$y = q_1 + b_1 q_1^2 + c_1 q_1^3 + d_1 q_1^4 + \ldots \tag{35}$$

From (17), by adding unity to each member, we have

$$1+q_1 = 1+y+\frac{y^2}{\underline{|2}}+\frac{y^3}{\underline{|3}}+\frac{y^4}{\underline{|4}}+\ .\ .\ .\ . = e^y \tag{36}$$

or

$$y = \log_e(1+q_1) \tag{37}$$

$$\therefore\ y = q_1 - \tfrac{1}{2}q_1^2 + \tfrac{1}{3}q_1^3 - \tfrac{1}{4}q_1^4 + \ .\ .\ .\ . \tag{38}$$

Comparing coefficients in (35) and (38), we find

$$b_1 = -\tfrac{1}{2} \qquad c_1 = +\tfrac{1}{3} \qquad d_1 = -\tfrac{1}{4} \tag{39}$$

Substituting these values in (34), we obtain

$$x_1 \equiv \omega F'(t) = \Delta_0' - \frac{\Delta_0''}{2} + \frac{\Delta_0'''}{3} - \frac{\Delta_0^{\text{iv}}}{4} + \ .\ .\ .\ . \tag{40}$$

Again, from (38), we derive

$$y^n = \left(q_1 - \frac{q_1^2}{2} + \frac{q_1^3}{3} - \frac{q_1^4}{4} + \ .\ .\ .\ .\right)^n \tag{41}$$

$$\therefore\ y^n = q_1^n - \frac{n}{2}q_1^{n+1} + \frac{n}{24}(3n+5)\,q_1^{n+2} - \frac{n}{48}(n+2)(n+3)\,q_1^{n+3} + \ .\ .\ .\ . \tag{42}$$

Equating coefficients in (33) and (42), we find

$$b_n = -\frac{n}{2}\ ,\quad c_n = +\frac{n}{24}(3n+5)\ ,\quad d_n = -\frac{n}{48}(n+2)(n+3)\ ; \tag{43}$$

These values being substituted in (30), the latter becomes

$$x_n \equiv \omega^n F^{(n)}(t) = \Delta_0^{(n)} - \frac{n}{2}\Delta_0^{(n+1)} + \frac{n}{24}(3n+5)\,\Delta_0^{(n+2)} - \frac{n}{48}(n+2)(n+3)\,\Delta_0^{(n+3)} + \ .\ . \tag{44}$$

Using the symbolic notation adopted in (21), we have the following expressions:

$$\left.\begin{aligned}
D &= \Delta_0 - \tfrac{1}{2}\Delta_0^2 + \tfrac{1}{3}\Delta_0^3 - \tfrac{1}{4}\Delta_0^4 + \tfrac{1}{5}\Delta_0^5 - \ .\ .\ . \\
D^2 &= (\Delta_0 - \tfrac{1}{2}\Delta_0^2 + \tfrac{1}{3}\Delta_0^3 - \tfrac{1}{4}\Delta_0^4 + \ .\ .\ .)^2 = \Delta_0^2 - \Delta_0^3 + \tfrac{11}{12}\Delta_0^4 - \tfrac{5}{6}\Delta_0^5 + \ .\ .\ . \\
D^3 &= (\Delta_0 - \tfrac{1}{2}\Delta_0^2 + \tfrac{1}{3}\Delta_0^3 - \tfrac{1}{4}\Delta_0^4 + \ .\ .\ .)^3 = \Delta_0^3 - \tfrac{3}{2}\Delta_0^4 + \tfrac{7}{4}\Delta_0^5 - \tfrac{15}{8}\Delta_0^6 + \ .\ .\ . \\
&\ldots\ldots\ldots\ldots\ldots\ldots\ldots\ldots\ldots\ldots \\
D^n &= (\Delta_0 - \tfrac{1}{2}\Delta_0^2 + \tfrac{1}{3}\Delta_0^3 - \tfrac{1}{4}\Delta_0^4 + \ .\ .\ .)^n \\
&= \Delta_0^n - \frac{n}{2}\Delta_0^{n+1} + \frac{n}{24}(3n+5)\,\Delta_0^{n+2} - \frac{n}{48}(n+2)(n+3)\,\Delta_0^{n+3} + \ .\ .\ .
\end{aligned}\right\} \tag{45}$$

19. *Effect of a Change in the Argument Interval* ω, *upon the Magnitude of the Several Orders of Differences.*—Let us now suppose

that a second tabulation of $F(T)$ has been made, differing from the first only in the value of the interval, ω. Let $\omega' = m\omega$ be the interval of the argument in the second table; denoting the differences by $\delta', \delta'', \delta''', \ldots.,$ the new table will run as follows:

T	$F(T)$	δ'	δ''	δ'''	δ^{iv}	. . .
t	$F(t)$					
$t+m\omega$	$F(t+m\omega)$	δ_0'				
$t+2m\omega$	$F(t+2m\omega)$	δ_1'	δ_0''			
$t+3m\omega$	$F(t+3m\omega)$	δ_2'	δ_1''	δ_0'''	δ_0^{iv}	. . .
$t+4m\omega$	$F(t+4m\omega)$	δ_3'	δ_2''	δ_1'''	δ_1^{iv}	. . .
				δ_2'''		
.	.	.	.	.	.	.
.	.	.	.	.	.	.
.	.	.	.	.	.	.

We proceed to investigate the relations between $\delta', \delta'', \delta''', \ldots\ldots$ and $\Delta', \Delta'', \Delta''', \ldots.$ No restriction is placed upon the value of m; in the applications of the resulting formulae, however, m will usually be regarded as a positive proper fraction. The second tabulation will then give the function for closer values of T than the first.

Since the value of ω is arbitrary, we may write $m\omega$ for ω in the right-hand member of (15), and thus obtain the expression for $\delta_0^{(n)}$; making this substitution, we find

$$\delta_0^{(n)} = m^n\omega^n F^{(n)}(t) + B_n m^{n+1}\omega^{n+1}F^{(n+1)}(t) + C_n m^{n+2}\omega^{n+2}F^{(n+2)}(t) + \ldots. \tag{46}$$

If, as above, we write x_r for $\omega^r F^{(r)}(t)$, this equation becomes

$$\delta_0^{(n)} = m^n x_n + B_n m^{n+1} x_{n+1} + C_n m^{n+2} x_{n+2} + \ldots. \tag{47}$$

From (30) we obtain, in succession,

$$\left.\begin{aligned} x_n &= \Delta_0^{(n)} + b_n\Delta_0^{(n+1)} + c_n\Delta_0^{(n+2)} + d_n\Delta_0^{(n+3)} + \ldots. \\ x_{n+1} &= \Delta_0^{(n+1)} + b_{n+1}\Delta_0^{(n+2)} + c_{n+1}\Delta_0^{(n+3)} + d_{n+1}\Delta_0^{(n+4)} + \ldots. \\ &\ldots\ldots\ldots\ldots\ldots\ldots \end{aligned}\right\} \tag{48}$$

Eliminating $x_n, x_{n+1}, \ldots.$ from (47), by means of (48), there results an equation of the form

$$\delta_0^{(n)} = m^n\Delta_0^{(n)} + \beta_n\Delta_0^{(n+1)} + \gamma_n\Delta_0^{(n+2)} + \ldots. \tag{49}$$

which, for $n = 1$, becomes

$$\delta_0' = m\Delta_0' + \beta_1\Delta_0'' + \gamma_1\Delta_0''' + \ldots. \tag{50}$$

Now let

$$z_n \equiv m^n y^n + B_n m^{n+1} y^{n+1} + C_n m^{n+2} y^{n+2} + \ldots \tag{51}$$

be an auxiliary expression, such that the coefficient of y^{n+r} is the coefficient of x_{n+r} in (47).

From (33) we obtain, in succession,

$$\left.\begin{aligned} y^n &= \varphi_1^n + b_n \varphi_1^{n+1} + c_n \varphi_1^{n+2} + d_n \varphi_1^{n+3} + \\ y^{n+1} &= \varphi_1^{n+1} + b_{n+1} \varphi_1^{n+2} + c_{n+1} \varphi_1^{n+3} + d_{n+1} \varphi_1^{n+4} + \quad \ldots \\ &\ldots\ldots\ldots\ldots\ldots\ldots\ldots\ldots \end{aligned}\right\} \tag{52}$$

Now, to eliminate $y^n, y^{n+1}, \ldots$ from (51), by means of (52), we must perform precisely the same algebraic steps as in the derivation of equation (49) from (47) and (48); we shall therefore obtain

$$z_n = m^n \varphi_1^n + \beta_n \varphi_1^{n+1} + \gamma_n \varphi_1^{n+2} + \ldots \tag{53}$$

and, for $n = 1$, we have

$$z_1 = m\varphi_1 + \beta_1 \varphi_1^2 + \gamma_1 \varphi_1^3 + \ldots \tag{54}$$

Now the equation (51) may be written

$$z_n = (my)^n + B_n (my)^{n+1} + C_n (my)^{n+2} + \ldots$$

Whence, by (15*a*), we have

$$z_n = \varphi_n(my) \tag{55}$$

and hence, also,

$$z_1 = \varphi_1(my)$$

or, by (17),

$$z_1 = (my) + \frac{(my)^2}{\underline{|2}} + \frac{(my)^3}{\underline{|3}} + \frac{(my)^4}{\underline{|4}} + \ldots = e^{my} - 1$$

$$\therefore 1 + z_1 = e^{my} \tag{56}$$

Also, from (36), we have

$$1 + \varphi_1 = e^y$$

the combination of which with (56) gives

$$1 + z_1 = (1+\varphi_1)^m = 1 + m\varphi_1 + \frac{m(m-1)}{\underline{|2}} \varphi_1^2 + \ldots + \frac{m(m-1)\ldots(m-r+1)}{\underline{|r}} \varphi_1^r + \ldots$$

or

$$z_1 = m\varphi_1 + \frac{m(m-1)}{\underline{|2}} \varphi_1^2 + \ldots + \frac{m(m-1)\ldots(m-r+1)}{\underline{|r}} \varphi_1^r + \ldots \tag{57}$$

Comparing (54) and (57), we find

$$\beta_1 = \frac{m(m-1)}{\underline{|2}} \quad , \quad \gamma_1 = \frac{m(m-1)(m-2)}{\underline{|3}} \quad , \tag{58}$$

Substituting these values in (50), we obtain the following fundamental relation:

$$\delta_0' = m\varDelta_0' + \frac{m(m-1)}{\underline{|2}}\varDelta_0'' + \ldots\ldots + \frac{m(m-1)\ldots\ldots(m-r+1)}{\underline{|r}}\varDelta_0^{(r)} + \ldots\ldots \tag{59}$$

Again, using the relation $q_n = q_1^n$, we obtain from (55),

$$z_n = q_n(my) = \{q_1(my)\}^n = z_1^n \tag{60}$$

Hence, from (57), we find

$$z_n = \left(mq_1 + \frac{m(m-1)}{\underline{|2}} q_1^2 + \frac{m(m-1)(m-2)}{\underline{|3}} q_1^3 + \ldots\ldots\right)^n$$

Expanding and factoring, we obtain

$$z_n = m^n q_1^n + \frac{n}{2} m^n (m-1) q_1^{n+1} + \frac{n}{24}\Big[(3n+1)m-(3n+5)\Big] m^n (m-1) q_1^{n+2}$$
$$+ \frac{n}{48}\Big[n(n+1)m^2 - 2(n^2+3n+1)m + (n+2)(n+3)\Big] m^n (m-1) q_1^{n+3} + \ldots \tag{61}$$

Equating coefficients of like powers of q_1 in (53) and (61), we have

$$\beta_n = \frac{n}{2} m^n (m-1) \quad , \quad \gamma_n = \frac{n}{24} m^n (m-1)\Big[(3n+1)m-(3n+5)\Big] \quad , \tag{62}$$

Substituting these values in (49), the latter becomes

$$\delta_0^{(n)} = m^n \varDelta_0^{(n)} + \frac{n}{2} m^n (m-1)\varDelta_0^{(n+1)} + \frac{n}{24} m^n (m-1)\Big[(3n+1)m-(3n+5)\Big]\varDelta_0^{(n+2)} \tag{63}$$
$$+ \frac{n}{48} m^n (m-1)\Big[n(n+1)m^2 - 2(n^2+3n+1)m + (n+2)(n+3)\Big]\varDelta_0^{(n+3)} + \ldots\ldots$$

Finally, we may symbolize these results by the following expressions: (64)

$$= m\varDelta_0 + \frac{m(m-1)}{\underline{|2}}\varDelta_0^2 + \frac{m(m-1)(m-2)}{\underline{|3}}\varDelta_0^3 + \frac{m(m-1)\ldots(m-3)}{\underline{|4}}\varDelta_0^4 + \frac{m(m-1)\ldots(m-4)}{\underline{|5}}\varDelta_0^5 + \ldots$$

$$^2 = \left(m\varDelta_0 + \frac{m(m-1)}{\underline{|2}}\varDelta_0^2 + \frac{m(m-1)(m-2)}{\underline{|3}}\varDelta_0^3 + \ldots\ldots\right)^2$$
$$= m^2\varDelta_0^2 + m^2(m-1)\varDelta_0^3 + \frac{m^2}{12}(m-1)(7m-11)\varDelta_0^4 + \frac{m^2}{12}(m-1)(m-2)(3m-5)\varDelta_0^5 +$$

$$^3 = \left(m\varDelta_0 + \frac{m(m-1)}{\underline{|2}}\varDelta_0^2 + \ldots\ldots\right)^3 = m^3\varDelta_0^3 + \tfrac{3}{2} m^3(m-1)\varDelta_0^4 + \frac{m^3}{4}(m-1)(5m-7)\varDelta_0^5 + \ldots\ldots$$

$$^4 = \left(m\varDelta_0 + \frac{m(m-1)}{\underline{|2}}\varDelta_0^2 + \ldots\ldots\right)^4 = m^4\varDelta_0^4 + 2m^4(m-1)\varDelta_0^5 + \frac{m^4}{6}(m-1)(13m-17)\varDelta_0^6 + \ldots\ldots$$

$$^5 = \left(m\varDelta_0 + \frac{m(m-1)}{\underline{|2}}\varDelta_0^2 + \ldots\ldots\right)^5 = m^5\varDelta_0^5 + \tfrac{5}{2} m^5(m-1)\varDelta_0^6 + \tfrac{5}{8} m^5(m-1)(4m-5)\varDelta_0^7 +$$

. . . .

20. THEOREM VII.—*If the nth differences of a given series of functions are numerically large as compared with all the following differences, then, if the series be re-tabulated with the argument interval m times its original value, the nth differences of the new series will be approximately m^n times the corresponding n^{th} differences of the original series.*

The theorem is a direct interpretation of equation (63). For, if $\Delta_0^{(n+1)}, \Delta_0^{(n+2)}, \ldots$ are all small in comparison with $\Delta_0^{(n)}$, then the approximate value of $\delta_0^{(n)}$ is $m^n \Delta_0^{(n)}$.

COROLLARY.—*If the nth differences of the given series are constant, then the nth differences of the new series are also constant, and equal to m^n times the original n^{th} differences.*

For, if $\Delta^{(n)}$ is constant, $\Delta^{(n+1)}, \Delta^{(n+2)}, \ldots$ are all zero, and hence (63) gives, rigorously,

$$\delta^{(n)} = m^n \Delta^{(n)}$$

21. To illustrate the foregoing results, we take the following table of cubes:

T	$F(T) \equiv T^3$	Δ'	Δ''	Δ'''
100	1000000			
		+ 92727		
103	1092727		+5562	
		98289		+162
106	1191016		5724	
		104013		162
109	1295029		5886	
		109899		+162
112	1404928		+6048	
		+115947		
115	1520875			

Here the interval $\omega = 3$. If we take $m = \frac{1}{3}$, the interval is reduced to 1, and hence the new table is as follows:

T	T^3	δ'	δ''	δ'''
100	1000000			
		+30301		
101	1030301		+606	
		30907		+6
102	1061208		612	
		31519		6
103	1092727		618	
		32137		+6
104	1124864		+624	
		+32761		
105	1157625			

We now test the first three of the equations (64); substituting in the latter $m=\frac{1}{3}$, and observing that the differences beyond Δ''' vanish, we find

$$\delta_0' = \tfrac{1}{3}\Delta_0' - \tfrac{1}{9}\Delta_0'' + \tfrac{5}{81}\Delta_0''' \quad , \quad \delta_0'' = \tfrac{1}{9}\Delta_0'' - \tfrac{2}{27}\Delta_0''' \quad , \quad \delta_0''' = \tfrac{1}{27}\Delta_0''' \tag{65}$$

From the first of the above tables, we take

$$\Delta_0' = +92727 \qquad \Delta_0'' = +5562 \qquad \Delta_0''' = +162$$

Whence, from (65), we derive

$$\delta_0' = 30909 - 618 + 10 = 30301 \qquad \delta_0'' = 618 - 12 = 606 \qquad \delta_0''' = 6$$

which agree exactly with the values found in the second table above. It will be observed that δ_0' and δ_0'' come within $\frac{1}{50}$ part of equaling $\frac{1}{3}\Delta_0'$ and $\frac{1}{9}\Delta_0''$, respectively; while $\delta_0''' = \frac{1}{27}\Delta_0'''$, exactly. These relations are in accord with Theorem VII.

22. *To Express the Differences of $F(T)$ in Terms of the Given Functions only.*—Let the given series be F_0, F_1, F_2, F_3, ... ; then the first differences are F_1-F_0, F_2-F_1, F_3-F_2, ... ; the second differences, $F_2-2F_1+F_0$, $F_3-2F_2+F_1$, ... ; the third differences, $F_3-3F_2+3F_1-F_0$, $F_4-3F_3+3F_2-F_1$, ... ; and so on. The coefficients evidently follow the *binomial law*. Thus we have generally

$$\Delta_0^{(n)} = F_n - nF_{n-1} + \frac{n(n-1)}{\lfloor 2} F_{n-2} - \ . \ . + (-1)^r {}_nC_r F_{n-r} \pm \ . \ . + (-1)^{n-1} nF_1 + (-1)^n F_0 \tag{66}$$

in which, according to the usual notation, we put ${}_nC_r$ for the coefficient of x^r in the expansion of $(1+x)^n$.

To prove (66), let us assume it true for the index n; then the expression for the nth difference immediately following $\Delta_0^{(n)}$ (*i.e.*, $\Delta_1^{(n)}$) will be obtained by increasing the subscripts of F_n, F_{n-1}, ... in (66) by unity. We therefore have

$$\Delta_1^{(n)} = F_{n+1} - nF_n + \frac{n(n-1)}{\lfloor 2} F_{n-1} - \ . \ . + (-1)^{r+1} {}_nC_{r+1} F_{n-r} \pm \ . \ . + (-1)^n F_1 \tag{67}$$

Subtracting (66) from (67), we find

$$\Delta_0^{(n+1)} = \Delta_1^{(n)} - \Delta_0^{(n)} = F_{n+1} - (n+1)F_n + \frac{(n+1)n}{\lfloor 2} F_{n-1} - \ . \ . \ . \ .$$
$$+ (-1)^{r+1} ({}_nC_{r+1} + {}_nC_r) F_{n-r} \pm \ . \ . \ . + (-1)^n (n+1) F_1 + (-1)^{n+1} F_0$$

But, as proved in Algebra, we have

$${}_nC_{r+1} + {}_nC_r = {}_{n+1}C_{r+1}$$

and hence the preceding equation becomes

$$\Delta_0^{(n+1)} = F_{n+1} - (n+1)F_n + \frac{(n+1)n}{\lfloor 2}F_{n-1} - \ldots + (-1)^{r+1}\,{}_{n+1}C_{r+1}F_{n-r} \pm \ldots + (-1)^{n+1}F_0 \qquad (68)$$

It follows from (68) that if the law expressed in (66) holds for n, it also holds for $n+1$. But we have seen above that the expression is true for $n=1$, 2 and 3. Hence it is true for $n=4$, and so on indefinitely; the equation (66) is therefore true for all positive integral values of n.

23. *To Express Any Function of a Given Series in Terms of Some Particular Function* (F_0), *and of the Differences* (a_0, b_0, c_0, ...) *which Follow that Function.*—As before, let F_0, F_1, F_2, F_3, ... denote the given series, the differences being taken as in the schedule below:

$F(T)$	Δ'	Δ''	Δ'''	Δ^{iv}	Δ^{v}	Δ^{vi}
F_0						
F_1	a_0	b_0				
F_2	a_1	b_1	c_0	d_0		
F_3	a_2	b_2	c_1	d_1	e_0	f_0
F_4	a_3	b_3	c_2	d_2	e_1	f_1
.	a_4	.	c_3	.	e_2	.
.	.	.	.	.	.	.
.	.	.	.	.	.	.
.	.	.	.	.	.	.
F_{n-1}	a_{n-1}	b_{n-2}	.	.		.
F_n	a_n	b_{n-1}		.		.
F_{n+1}	.	.		.		.
.	.	.		.		.
.	.	.	.	.	.	.

Let it be required to express F_n in terms of F_0, a_0, b_0, c_0, d_0, ... From the nature of the differences, we have

$$\begin{aligned}
F_1 &= F_0 + a_0 \\
F_2 &= F_1 + a_1 = (F_0+a_0) + (a_0+b_0) = F_0 + 2a_0 + b_0 \\
F_3 &= F_2 + a_2 = (F_0+2a_0+b_0) + (a_0+2b_0+c_0) = F_0 + 3a_0 + 3b_0 + c_0
\end{aligned}$$

and so on. The coefficients again follow the binomial law, which suggests for the form of the general term —

$$F_n = F_0 + na_0 + \frac{n(n-1)}{\underline{|2}} b_0 + \frac{n(n-1)(n-2)}{\underline{|3}} c_0 + \dots \qquad (69)$$

To prove (69) by induction, we assume that it is true for the index n. Moreover, we evidently have

$$F_{n+1} = F_n + a_n$$

We may now find a_n in terms of $a_0, b_0, c_0, d_0, \dots$ from (69), — since the relation is here the same as the relation of F_n to $F_0, a_0, b_0, c_0, \dots$; thus we obtain

$$a_n = a_0 + nb_0 + \frac{n(n-1)}{\underline{|2}} c_0 + \dots$$

Adding this value of a_n to that of F_n given by (69), we find*

$$F_{n+1} = F_n + a_n = F_0 + (n+1)a_0 + \frac{(n+1)n}{\underline{|2}} b_0 + \frac{(n+1)n(n-1)}{\underline{|3}} c_0 + \dots \qquad (70)$$

Thus, having assumed the relation (69) to be true for the index n, we find by (70) that it is also true when $n+1$ is written for n; but we have shown directly that (69) holds for $n = 1$, 2 and 3. The formula (69) is therefore true for all positive integral values of n.

* We here omit the proof for the general term, since the process is the same as in § 22.

EXAMPLES.

1. Tabulate the five-place logarithms of 25, 30, 35, ... 65, 70, and take the differences to the fifth order inclusive. Retain a copy of the table for further use.

2. Tabulate $F(T) \equiv \log\cos T$, to five decimals, for $T = 50°$, $53°, 56°, \ldots. 74°, 77°$; difference to the fifth order, as in Example 1. Retain a copy of the table.

3. Verify the accuracy of both the functions and their differences in Examples 1 and 2, by noting the degree of regularity in Δ^{v}, according to the method of §8.

4. Also, rigorously check the differencing in the above examples, by taking the algebraic sum of each separate order, as explained in §3.

5. Add the two series of functions tabulated in Examples 1 and 2; difference the new series as before, and see that the resulting values of Δ^{v} are the sums of the fifth differences of the other series, according to Theorem IV.

6. Correct the errors in the following tables by the method of differences:

(a)

T	$F(T) \equiv \frac{1}{T}$
0.21	4.762
.23	4.348
.25	4.000
.27	3.704
.29	3.465
.31	3.226
.33	3.030
.35	2.857
.37	2.703
0.39	2.564

(b)

Appa. Alt. of Star	Mean Refraction
°	′ ″
10	5 19.2
12	4 27.5
14	3 49.5
16	3 18.4
18	2 57.5
20	2 38.8
22	2 23.3
24	2 10.2
26	1 58.9

(c)

Latitude	Reduction
°	′ ″
0	0 0.00
2	0 48.02
4	1 35.80
6	2 23.12
8	3 9.75
10	3 55.11
12	4 40.05
14	5 23.28
16	6 4.95
18	6 44.86

(*d*)

T	$F(T) \equiv T^{\sin T}$
0.48	0.7125
.50	.7173
.52	.7226
.54	.7273
.56	.7349
.58	.7419
.60	.7494
.62	.7568
.64	.7660
.66	.7751
.68	.7847
.70	.7947
0.72	0.8052

(*e*)

Date 1898	Log. Dist. of *Mars from Earth*
Sept. 17	0.139162
21	.130819
25	.122145
29	.113130
Oct. 3	.103759
7	.094015
11	.083857
15	.073360
19	.062478
23	.051135
27	.039438
31	.027351
Nov. 4	0.014875

(*f*)

Date 1898	Lunar Dist. of *Jupiter* ° ′ ″
Dec. 1.0	105 5 59
1.5	99 18 28
2.0	93 31 31
2.5	87 44 46
3.0	81 57 48
3.5	76 10 17
4.0	70 21 14
4.5	64 30 37
5.0	58 39 44
5.5	52 42 5
6.0	46 43 12
6.5	40 40 43
7.0	34 34 29

7. Tabulate the following rational integral functions for the assigned values of the argument. Before taking the differences, state at which order the latter become constant, and compute the constant value in each case, by Theorem V. Then take the differences, and see that the results agree with the computed values.

(*a*) $F(T) \equiv T^6 - 50T^4 + 100T^2$.
(Tabulate for $T = -8, -6, -4, -2, 0, +2, +4, +6, +8$.)

(*b*) $F(T) \equiv 2T^3 - 7T - 400$. ($T = 8.0, 8.3, 8.6, \ldots 9.8$.)

(*c*) $F(T) \equiv 0.16T^4 - 0.3T^3$. ($T = 2, 3, 4, 5, 6, 7, 8$.)

8. By means of the first of equations (1), compute the value of Δ' which immediately follows log cos 56° in the table of Example 2. The value of ω ($= 3°$) must be expressed in circular measure. Compare the computed with the tabular value.

9. Tabulate $F(T) \equiv \log T$, to five places of decimals, for $T = 30, 40, 50, 60, 70$; denote this table by B, and that of Example 1 by A. A and B then differ only in ω, the interval having now been doubled. Then, in the second of the equations (64), put $m = 2$, and substitute from A the values of Δ_0'', Δ_0''', Δ_0^{iv}, and Δ_0^{v}, which correspond to $T = 40$. Whence, compute the value of δ_0'' corresponding to $T = 40$ in B, and compare computed with actual value.

10. In Example 1, compute the quantities Δ_0^{iv} and $F_5 (= \log 50)$, by (66) and (69) respectively; compare the results with the values found in the table.

CHAPTER II.

OF INTERPOLATION.

24. *Statement of the Problem.*—Given a series of numerical values of a function, for equidistant values of the argument, it is required to find the value of the function for *any intermediate* value of the argument, independently of the analytical form of the function, which may or may not be given.

Interpolation is the process or method by which the required values are found.

Without certain restrictions or assumptions as to the character of the function and the interval of its tabulation, the problem of interpolation is an indeterminate one. Thus it is evident, *a priori*, that from a series of temperatures recorded for every noon at a given station, it would be impossible to obtain by interpolation the temperature at 8.00 P.M., for a given day. If, *per contra*, the thermometric readings were recorded for 7.00, 7.10, 7.20, 7.30, ... P.M., it is highly probable that the temperature at 7.14 P.M. could be interpolated with accuracy.

The *Nautical Almanac* gives the heliocentric longitude of *Jupiter* for every 4th day; but, because of the slow, continuous, and systematic character of *Jupiter's* orbital motion, it is found sufficient to compute the longitudes from the tables direct for every 40th day only. The intermediate places are then readily interpolated with an accuracy which equals, if indeed it does not exceed, that of direct computation.

The moon's longitude is given in the *Nautical Almanac* for every twelve hours; for the moon's orbital motion is so rapid and complicated that it would prove inexpedient to attempt the interpolation of accurate values of the longitude from an ephemeris given for whole day intervals.

It therefore appears that, to render the problem of interpolation determinate, the tabular interval (ω) must be sufficiently small that the nature or law of the function will be definitively shown by the tabular values in question. The condition thus imposed will be satisfied when, in a given table, the differences become either *rigorously* or *sensibly* constant at some particular order.* This follows from the fact, soon to be proved, that for all such cases a formula of interpolation can be established, either *rigorously* or *sensibly* true, according to the foregoing distinction.

25. *Extension of Formula* (69) *to Fractional and Negative Values of* n, *Provided the Differences of Some Particular Order are Constant.*—We have shown (Theorem V) that the differences of a rational integral function vanish beyond a certain order. We proceed to prove that, for any such function, the formula (69) is rigorously true for *all* values of n.

Let $F(T)$ denote any function whose differences become constant at the order i, and let $\Delta^{(i)} = l_0$; $F(T)$ and its differences are then shown in the schedule on following page.

---

*Excepting, of course, any *periodic* function whose tabular interval (ω) differs but little from some multiple of its period, P. An example of such a series is the following:

Date, 1898	Day of the Year	Heliocentric Longitude of *Mercury*	Δ'	Δ''	Δ'''
		° ′	° ′	′	′
Jan. 4	4	93 0			
			+12 33		
Apr. 4	94	105 33		−26	
			12 7		−7
July 3	184	117 40		33	
			11 34		−5
Oct. 1	274	129 14		−38	
			+10 56		
Dec. 30	364	140 10			

where P (the time of one revolution of *Mercury*) $= 87.97$ days; and hence $\omega = 90^d = P + 2^d.03$. The differences Δ' therefore correspond to a tabular interval of 2.03 days, and *not* to the interval 90 days, as the table itself would indicate. Now, the actual value of *Mercury's* longitude for Jan. 14 is found from the *Nautical Almanac* to be $l = 149° 40'$; if, however, we fail to account for the periodic character of this function, and argue solely from the numerical data at hand, we find by a rough interpolation, for Jan. 14,

$$l = 93°.0 + (\tfrac{10}{90} \times 12°.6) = 94°.4$$

which bears no relation to the truth. The possibility of thus committing serious error through failing to account for completed periods or revolutions, suggests the necessity of caution in this direction.

T	$F(T)$	Δ'	Δ''	. . .	$\Delta^{(i)}$
t	F_0				
$t+\omega$	F_1	a_0	b_0		
$t+2\omega$	F_2	a_1	b_1	. . .	l_0
$t+3\omega$	F_3	a_2	b_2	.	l_0
.	.	.	.	.	l_0
.	.	.	.	.	l_0
.	.	.	.	.	
$t+(i+2)\omega$	F_{i+2}	.	b_{i+1}	. . .	
$t+(i+3)\omega$	F_{i+3}	a_{i+2}			

From (30) we obtain, in succession,

$$\omega^i F^{(i)}(t) = \Delta_0^{(i)} + b_i \Delta_0^{(i+1)} + c_i \Delta_0^{(i+2)} + \ldots .$$
$$\omega^{i+1} F^{(i+1)}(t) = \Delta_0^{(i+1)} + b_{i+1} \Delta_0^{(i+2)} + \ldots .$$
$$\omega^{i+2} F^{(i+2)}(t) = \Delta_0^{(i+2)} + \ldots .$$
$$\ldots\ldots\ldots\ldots\ldots\ldots\ldots\ldots\ldots$$

With the condition assumed, these equations give

$$\omega^i F^{(i)}(t) = l_0$$
$$\omega^{i+1} F^{(i+1)}(t) = \omega^{i+2} F^{(i+2)}(t) = \ldots . = 0$$

Hence, in this case, the expansions (0) end at the $(i+1)$th term. It follows that, under the present assumption, the expansions (0) are valid; in other words, $F(t+n\omega)$ is capable of expansion by TAYLOR'S Theorem for all values of n within the limits of the given table. Hence, for *all* such values, we have

$$F_n \equiv F(t+n\omega) = F(t) + n\omega F'(t) + \frac{n^2\omega^2}{\underline{|2}} F''(t) + \ldots . + \frac{n^i\omega^i}{\underline{|i}} F^{(i)}(t) \qquad (71)$$

Let us now consider the expression

$$Q \equiv F_0 + na_0 + \frac{n(n-1)}{\underline{|2}} b_0 + \ldots . + \frac{n(n-1)\ \ldots\ (n-i+1)}{\underline{|i}} l_0 \qquad (72)$$

Substituting, successively, $n = 0, 1, 2, 3, \ldots . i+3$, in (72), we get, according to (69),

$$Q = F_0,\ F_1,\ F_2\ F_3, \ldots . F_{i+3}, \text{ respectively.}$$

Substituting these same values of n in (71), we evidently obtain the same results, namely—

$$F_n = F_0,\ F_1,\ F_2,\ F_3, \ldots . F_{i+3}, \text{ in succession.}$$

Hence, F_n and Q are equal to each other for more than i values of n. But F_n and Q are both expressions of the degree i in n. Now, when two expressions of the degree i in n are equal to each other for more than i values of n, they are equal for all values of n. Therefore, for *all* values of n, fractional and negative, we have

$$F_n \equiv F(t+n\omega) = F_0 + na_0 + \frac{n(n-1)}{\underline{|2}} b_0 + \dots + \frac{n(n-1)\dots(n-i+1)}{\underline{|i}} l_0 \quad (73)$$

provided that $\varDelta^{(i)} = l_0 =$ constant. This is the fundamental formula of interpolation, and is known as NEWTON'S *Formula*.

26. *Second Proof of* NEWTON'S *Formula, for Constant Values of* $\varDelta^{(i)}$.—Formula (73) is readily proved by means of equation (59), in which m may have any value. The only condition necessary for the validity of (59) is that the expansions (0) are themselves valid. But since we assume that the differences beyond $\varDelta^{(i)}$ vanish, it follows (as proved in the last section) that the expansions (0) *are* valid. Hence (59) gives, rigorously,

$$\delta_0' = m\varDelta_0' + \frac{m(m-1)}{\underline{|2}}\varDelta_0'' + \dots + \frac{m(m-1)\dots(m-i+1)}{\underline{|i}}\varDelta_0^{(i)}$$

From the definition of δ_0' (see schedule, p. 31), we have

$$\delta_0' = F(t+m\omega) - F(t) = F_m - F_0$$

$$\therefore \quad F_m \equiv F(t+m\omega) = F_0 + \delta_0'$$

$$= F_0 + m\varDelta_0' + \frac{m(m-1)}{\underline{|2}}\varDelta_0'' + \dots + \frac{m(m-1)\dots(m-i+1)}{\underline{|i}}\varDelta_0^{(i)}$$

which is the same as formula (73), except that m is written for n.

27. *To Find n, the Interval of Interpolation.*—The binomial coefficients of NEWTON'S Formula are given in Table I, for every hundredth part of a unit in the argument n. The quantity n is called the *interval of interpolation*, and in practice is always less than unity. To obtain an expression for n, suppose that we are to interpolate the value of the function corresponding to the argument T, whose value lies between t and $t+\omega$; then we shall have

$$F_n \equiv F(t+n\omega) = F(T) \quad , \text{ or } \quad t+n\omega = T$$

and therefore

$$n = \frac{T-t}{\omega} \quad (74)$$

which determines the interval n.

28. Example.—From the following table of T^4, find the value of $(2.8)^4$ by Newton's Formula:

T	$F(T) \equiv T^4$	Δ'	Δ''	Δ'''	Δ^{iv}	Δ^{v}
2	16					
		+ 240				
4	256		+ 800			
		1040		+ 960		
6	1296		1760		+384	
		2800		1344		0
8	4096		3104		384	
		5904		1728		0
10	10000		4832		+384	
		10736		+2112		
12	20736		+6944			
		+17680				
14	38416					

Here we have

$$T = 2.8 \qquad a_0 = +240$$
$$t = 2 \qquad b_0 = +800$$
$$\omega = 2 \qquad c_0 = +960$$
$$n = \tfrac{2.8-2}{2} = 0.4 \qquad d_0 = +384$$
$$F_0 = 16 \qquad e_0 = 0$$

It will be convenient to denote the coefficients of $a_0, b_0, c_0, \ldots.$ in (73) by $A, B, C, \ldots.,$ respectively. Then, from Table I (with argument $n = 0.40$), or by direct computation, we find

$$A = +0.40 \qquad C = +0.0640$$
$$B = -0.12 \qquad D = -0.0416$$

We therefore obtain

$$\begin{aligned} F_0 &= +16.00 \\ Aa_0 &= +96.00 \\ Bb_0 &= -96.00 \\ Cc_0 &= +61.44 \\ Dd_0 &= -15.9744 \\ \hline \therefore\ (2.8)^4 = F_{0.4} &= +61.4656 \end{aligned}$$

This result is easily verified, and found exact to the last figure. However, since Table I does not in general give the exact mathematical values of the interpolating coefficients, it follows that functions interpolated in this manner cannot always be *absolutely* correct. The results may be, as in logarithmic computation, but close approximations to the truth.

29. *Backward Interpolation.*—When the interval of interpolation approaches unity, it is usually more convenient to proceed *backwards* from the function which *follows* the value sought. The problem,

therefore, is to find F_{-n}; for this purpose, let $F(T)$ be differenced as in the schedule below — the values of $\Delta^{(n)}$ being supposed constant as before:

T	$F(T)$	Δ'	Δ''	Δ'''	Δ^{iv}	. . .	$\Delta^{(n)}$
$t-3\omega$	F_{-3}		b_{-4}		d_{-5}	. . .	l_0
$t-2\omega$	F_{-2}	a_{-3}	b_{-3}	c_{-4}	d_{-4}	. . .	l_0
$t-\omega$	F_{-1}	a_{-2}	b_{-2}	c_{-3}	d_{-3}	. . .	l_0
t	F_0	a_{-1}	b_{-1}	c_{-2}	d_{-2}	. . .	l_0
$t+\omega$	F_1	a_0	b_0	c_{-1}	d_{-1}	. . .	l_0
$t+2\omega$	F_2	a_1	b_1	c_0	d_0	. . .	l_0
$t+3\omega$	F_3	a_2	b_2	c_1	d_1	. . .	l_0

We might substitute $-n$ for n in (73), and find directly,

$$F_{-n} = F_0 + (-n)a_0 + \frac{(-n)(-n-1)}{\underline{|2}}b_0 + \frac{(-n)(-n-1)(-n-2)}{\underline{|3}}c_0 + \ldots\ldots$$

But this formula, while true, is inconvenient from the fact that its coefficients neither converge as rapidly as the binomial coefficients for $+n$, nor can their numerical values be taken from Table I. To avoid the negative interval, we have only to suppose the series inverted, thus making F_3 the first, and F_{-3} the last of the tabular functions. Then, by Theorem III, the signs of Δ', Δ''', Δ^{v}, ... are changed, while the signs of Δ'', Δ^{iv}, ... are unaltered. Now the value of F_{-n} is obtained by interpolating *forward* with the interval $+n$ in the *inverted* series; hence the differences to be used in NEWTON'S Formula are—

$$-a_{-1},\ +b_{-2},\ -c_{-3},\ +d_{-4},\ \ldots\ldots$$

We therefore have, by (73),

$$F_{-n} \equiv F(t-n\omega) = F_0 - na_{-1} + \frac{n(n-1)}{\underline{|2}}b_{-2} - \frac{n(n-1)(n-2)}{\underline{|3}}c_{-3} + \frac{n(n-1)(n-2)(n-3)}{\underline{|4}}d_{-4} - \ldots \qquad (75)$$

the differences being taken as in the above schedule. The coefficients, as before, are taken from Table I with the argument n.

An immediate and important application of (75) is in finding the value of a function near the end of a given series. Thus, in the preceding schedule, suppose the series ended with F_0, and it were required to interpolate a value of F between F_{-1} and F_0: since the differences b_{-1}, c_{-1}, d_{-1}, ... (required in interpolating forward from F_{-1}) are not given in this case, the formula (75) must be used; n being the interval of the required function from F_0 toward F_{-1}.

EXAMPLE.— From the table of T^4 given on page 44, find the value of $(13.26)^4$.

Taking $t = 14$, we find

$$n = \frac{14 - 13.26}{2} = 0.37$$

which is the interval counted *backwards* from $F = 38416$. Hence, from Table I, we obtain

$$A = +0.37 \qquad C = +0.06333$$
$$B = -0.11655 \qquad D = -0.04164$$

And for the differences required by (75), we have

$$a_{-1} = +17680 \qquad c_{-3} = +2112$$
$$b_{-2} = +\ 6944 \qquad d_{-4} = +\ 384$$

Therefore, by (75), we derive

$$\begin{aligned} F_0 &= +38416.00 \\ -Aa_{-1} &= -\ 6541.60 \\ +Bb_{-2} &= -\ \ 809.32 \\ -Cc_{-3} &= -\ \ 133.75 \\ +Dd_{-4} &= -\ \ \ 15.99 \\ \hline \therefore F_n = (13.26)^4 &= +30915.34 \end{aligned}$$

By direct calculation, we find

$$(13.26)^4 = 30915.34492+$$

30. *Application of* NEWTON'S *Formula, when the Differences Become only Approximately Constant.*—We have proved (§§25 and 26) that (73) is true for all values of n, provided the differences of some particular order are *rigorously* constant. We now propose to show that, if the value of n lies between 0 and $+1$, the formula is very approximately true for the more frequent case in which the differences of some order become *approximately*, but not absolutely constant. The example given on page 8 is typical of this case; the numbers involved are not the true mathematical values of the quantities represented, and hence the irregularities, as already explained.

Let $F_0, F_1, F_2, F_3, \ldots . F_r, \ldots .$ denote a series of approximate tabular values of any function $F(T)$, given for equidistant values of T, and true to the *nearest* unit of their last figure; let $\bar{F}_0, \bar{F}_1, \bar{F}_2, \bar{F}_3, \ldots \bar{F}_r, \ldots$ denote the corresponding true mathematical values of the series, which we shall designate generally as $\bar{F}$; also, let $F_r = \bar{F}_r + f_r$; f_r being the difference between the true and approximate values, due to the omission of decimals in the tabular quantities.

The differences of $\bar{F}$, and those of the series $f_0, f_1, f_2, f_3, \ldots$, are now defined by the two schedules below:

T	$\bar{F}(T)$	Δ'	Δ''	Δ'''	. .	$\Delta^{(i)}$	$\Delta^{(i+1)}$	. .
t	$\bar{F}_0$							
$t+\omega$	$\bar{F}_1$	a_0						
$t+2\omega$	$\bar{F}_2$	a_1	b_0					
$t+3\omega$	$\bar{F}_3$	a_2	b_1	c_0	. .			
$t+4\omega$	$\bar{F}_4$	a_3	b_2	c_1	. .	l_0	m_0	
$t+5\omega$	$\bar{F}_5$	a_4	b_3	c_2	. .	l_1	m_1	. .
. . .	. . .	. .	b_4	c_3	. .	l_2	m_2	. .

(A)

T	f	Δ'	Δ''	Δ'''	. .	$\Delta^{(i)}$	$\Delta^{(i+1)}$	. .
t	f_0							
$t+\omega$	f_1	α_0						
$t+2\omega$	f_2	α_1	β_0					
$t+3\omega$	f_3	α_2	β_1	γ_0	. .			
$t+4\omega$	f_4	α_3	β_2	γ_1	. .	λ_0	μ_0	
$t+5\omega$	f_5	α_4	β_3	γ_2	. .	λ_1	μ_1	. .
. . .			β_4	γ_3	. .	λ_2	μ_2	. .

(B)

Then, since $F = \bar{F} + f$, it follows from Theorem IV that the differences of F are as given in the appended table:

T	$F(T)$	Δ'	Δ''	Δ'''	. .	$\Delta^{(i)}$	$\Delta^{(i+1)}$	. .
t	$F_0 = \bar{F}_0 + f_0$							
$t+\omega$	$F_1 = \bar{F}_1 + f_1$	$a_0+\alpha_0$						
$t+2\omega$	$F_2 = \bar{F}_2 + f_2$	$a_1+\alpha_1$	$b_0+\beta_0$					
$t+3\omega$	$F_3 = \bar{F}_3 + f_3$	$a_2+\alpha_2$	$b_1+\beta_1$	$c_0+\gamma_0$	. .			
$t+4\omega$	$F_4 = \bar{F}_4 + f_4$	$a_3+\alpha_3$	$b_2+\beta_2$	$c_1+\gamma_1$	. .	$l_0+\lambda_0$	$m_0+\mu_0$	. .
$t+5\omega$	$F_5 = \bar{F}_5 + f_5$	$a_4+\alpha_4$	$b_3+\beta_3$	$c_2+\gamma_2$	. .	$l_1+\lambda_1$	$m_1+\mu_1$	. .
. . .	. . . . .	. . .	$b_4+\beta_4$	$c_3+\gamma_3$	. .	$l_2+\lambda_2$	$m_2+\mu_2$	. .

(C)

Let us now suppose that the differences $\Delta^{(i+1)}$ in Table (C) are either alternately $+$ and $-$, or that $+$ and $-$ signs follow each other irregularly. Moreover, the foregoing definition of F requires that the terms in $\Delta^{(i+1)}$ are sufficiently small to indicate that no errors exceeding half a unit in the last place exist in the functions $F(T)$. The values of $\Delta^{(i)}$ are then approximately constant, and therefore Table (C) represents the typical case in practice. We proceed to investigate the accuracy of NEWTON's Formula as applied in this case; assuming that n is always taken within the limits 0 and +1, and that terms beyond $\Delta^{(i)}$ are neglected.

Applying (73) to find F_n from Table (C), and omitting the terms beyond $\Delta^{(i)}$, we have

$$F_n = (\overline{F}_0 + f_0) + A(a_0 + \alpha_0) + B(b_0 + \beta_0) + C(c_0 + \gamma_0) + \ldots\ldots + L(l_0 + \lambda_0) \qquad (76)$$

in which $A, B, C, \ldots\ldots L$ denote the binomial coefficients of the nth order. Let us now examine the approximate formula (76), to discover its *maximum* error *when all conditions conspire to that end.*

The formula (76) may be written

$$F_n = (\overline{F}_0 + Aa_0 + Bb_0 + \ldots\ldots + Ll_0) + (f_0 + A\alpha_0 + B\beta_0 + \ldots\ldots + L\lambda_0) \qquad (77)$$

For brevity, let us put

$$\left.\begin{aligned} Q &\equiv \overline{F}_0 + Aa_0 + Bb_0 + \ldots\ldots + Ll_0 \\ R &\equiv f_0 + A\alpha_0 + B\beta_0 + \ldots\ldots + L\lambda_0 \\ \therefore\ F_n &= Q + R \end{aligned}\right\} \qquad (77a)$$

It will be observed that Q is the value obtained for $\overline{F}_n$ when (73) is applied to Table (A), terms beyond $\Delta^{(i)}$ being neglected. We leave the discussion of Q for the present, to consider the quantity R, which evidently expresses the error of interpolation due to the unavoidable errors, f, contained in the tabular functions F.

Applying the formulae of §22 to the differences of Table (B), we have

$$\left.\begin{aligned} \alpha_0 &= f_1 - f_0 \\ \beta_0 &= f_2 - 2f_1 + f_0 \\ \gamma_0 &= f_3 - 3f_2 + 3f_1 - f_0 \\ \delta_0 &= f_4 - 4f_3 + 6f_2 - 4f_1 + f_0 \\ \epsilon_0 &= f_5 - 5f_4 + 10f_3 - 10f_2 + 5f_1 - f_0 \\ &\ldots\ldots\ldots\ldots\ldots\ldots \end{aligned}\right\} \qquad (78)$$

Hence, from (77*a*), we obtain

$$\begin{aligned} R &= f_0 + A\alpha_0 + B\beta_0 + C\gamma_0 + D\delta_0 + E\epsilon_0 + \ldots + L\lambda_0 \\ &= f_0 + A(f_1 - f_0) + B(f_2 - 2f_1 + f_0) + C(f_3 - 3f_2 + 3f_1 - f_0) \\ &\qquad + D(f_4 - 4f_3 + 6f_2 - 4f_1 + f_0) + \ldots \end{aligned}$$

$$\left.\begin{aligned} \therefore R = f_0(1 - A + B - C + D - E + \ldots \pm L) + f_1(A - 2B + 3C - 4D + 5E - \ldots) \\ + f_2(B - 3C + 6D - 10E + \ldots) + f_3(C - 4D + 10E - \ldots) \\ + f_4(D - 5E + \ldots) + f_5(E - \ldots) + \ldots \end{aligned}\right\} \quad (79)$$

Now the binomial coefficients $A, B, C, \ldots$ are connected by the following relations:

$$A = n \quad , \quad B = \left(\frac{n-1}{2}\right)A \quad , \quad C = \left(\frac{n-2}{3}\right)B \quad ,$$

Hence, since we have assumed that n lies between 0 and $+1$, it follows that $A, B, C, \ldots$ are alternately positive and negative, thus:

A	B	C	D	E	.	.
+	−	+	−	+	.	.

We therefore draw the following conclusions respecting (79):

The coefficient of f_1 is $+$;
" " " f_2 " $-$;
" " " f_3 " $+$;
" " " f_4 " $-$;
. . . .

Now, since the values of F are supposed true to the *nearest* unit of the last decimal figure, the quantities f may have any value between —0.5 and +0.5, in terms of the same unit; hence, it follows from the foregoing conclusions that if we take

$$f_1 = +0.5 \qquad f_2 = -0.5 \qquad f_3 = +0.5 \qquad f_4 = -0.5 \qquad \ldots \qquad (80)$$

the sum of all the terms *after the first* in the right-hand member of (79) will be numerically a *maximum*, with the $+$ sign.

We shall now show that the coefficient of f_0 in (79) is a positive number. For this pupose, let us consider the identity

$$(1-x)^{-1}(1-x)^n \equiv (1-x)^{n-1}$$

which, for all values of x numerically less than unity, may be expanded into the form

$$(1 + x + x^2 + x^3 + \ldots + x^i + \ldots)(1 - Ax + Bx^2 - Cx^3 + \ldots \pm Lx^i \mp \ldots) \equiv (1-x)^{n-1}$$

Upon equating the coefficients of x^i in the two members of this identity, we find

$$1-A+B-C+\ .\ .\ .\ \pm L = (-1)^i\ .\ \frac{(n-1)(n-2)(n-3)\ .\ .\ .\ (n-i)}{\underline{|i}}$$
$$= \left(1-\frac{n}{1}\right)\left(1-\frac{n}{2}\right)\left(1-\frac{n}{3}\right)\ .\ .\ .\ \left(1-\frac{n}{i}\right)$$

Now, the first member of this equation is the coefficient of f_0 in (79); and since the final member contains only positive factors, it follows that the coefficient of f_0 in (79) is a *positive* quantity. Accordingly, if we take $f_0 = +0.5$, in conjunction with the values of $f_1, f_2, f_3, \ldots\ldots$ designated in (80), the value of R given by (79) will then be the greatest possible under the assigned conditions.

We now append a table of the quantities $f_0, f_1, f_2, f_3, \ldots\ldots$ as above determined, with their differences:

T	f	Δ'	Δ''	Δ'''	Δ^{iv}	Δ^{v}	Δ^{vi}	Δ^{vii}
t	**+0.5**	**0.0**	**−1**	**+3**	**−7**	**+15**	**−31**	**+63**
$t+\omega$	+0.5	−1.0	+2	−4	+8	−16	+32	−64
$t+2\omega$	−0.5	+1.0	−2	+4	−8	+16	−32	. .
$t+3\omega$	+0.5	−1.0	+2	−4	+8	. .	. .	
$t+4\omega$	−0.5	+1.0	−2	. .	. .			
$t+5\omega$	+0.5	. .	. .					
. . .	. .							

(B′)

The special values which must be assigned to the quantities $f_0, \alpha_0, \beta_0, \gamma_0, \ldots.$ of Table (B) are, therefore,

f_0	α_0	β_0	γ_0	δ_0	ϵ_0
+0.5	0.0	−1	+3	−7	+15

in units of the last place of the tabular quantities F. Substituting these values in the original expression for R given in (77*a*), namely,

$$R = f_0 + A\alpha_0 + B\beta_0 + C\gamma_0 + \ .\ .\ .\ .$$

we obtain

$$R = +0.5 - B + 3C - 7D + 15E - 31F + 63G - \ .\ .\ .\ . \tag{81}$$

which gives the *maximum* value possible to R for $n \gtrless {}^{0}_{1}$.

To evaluate (81) for different values of n between 0 and $+1$, we make use of the following abridged table:

$n = A$	B	C	D	E	F	G
+	−	+	−	+	−	+
0.00	.0000	.0000	.0000	.0000	.0000	.0000
0.10	.0450	.0285	.0207	.0161	.0132	.0111
0.20	.0800	.0480	.0336	.0255	.0204	.0169
0.30	.1050	.0595	.0402	.0297	.0233	.0190
0.40	.1200	.0640	.0416	.0300	.0230	.0184
0.50	.1250	.0625	.0391	.0273	.0205	.0161
0.60	.1200	.0560	.0336	.0228	.0168	.0129
0.70	.1050	.0455	.0262	.0173	.0124	.0094
0.80	.0800	.0320	.0176	.0113	.0079	.0059
0.90	.0450	.0165	.0087	.0054	.0037	.0027
1.00	.0000	.0000	.0000	.0000	.0000	.0000
+	−	+	−	+	−	+

(D)

From these values we tabulate as follows:

n	**0.00**	**0.10**	**0.20**	**0.30**	**0.40**	**0.50**	**0.60**	**0.70**	**0.80**	**0.90**	**1.00**
		+	+	+	+	+	+	+	+	+	
$-\ B$	.000	.045	.080	.105	.120	.125	.120	.105	.080	.045	.000
$+\ 3C$	.000	.085	.144	.178	.192	.187	.168	.136	.096	.049	.000
$-\ 7D$	.000	.145	.235	.281	.291	.274	.235	.183	.123	.061	.000
$+15E$	.000	.241	.382	.445	.450	.409	.342	.259	.169	.081	.000
$-31F$	.000	.409	.632	.722	.713	.635	.521	.384	.245	.115	.000
$+63G$	.000	.699	1.065	1.197	1.159	1.014	.813	.592	.372	.170	.000

If, now, we let R_2, R_3, R_4, ... denote the values of R when differences beyond the 2d, 3d, 4th, ... order respectively are neglected, then, from (81), we find

$$\left.\begin{aligned} R_2 &= 0.5 - B \\ R_3 &= 0.5 - B + 3C \\ R_4 &= 0.5 - B + 3C - 7D \\ &\ldots\ldots\ldots\ldots \end{aligned}\right\} \qquad (82)$$

From the last table we obtain, by successive additions, the values of R_2, R_3, R_4, ... as defined by (82); these values are tabulated below:

n	**0.00**	**0.10**	**0.20**	**0.30**	**0.40**	**0.50**	**0.60**	**0.70**	**0.80**	**0.90**	**1.00**
R_2	0.50	0.55	0.58	0.60	0.62	0.63	0.62	0.60	0.58	0.55	0.50
R_3	0.50	0.63	0.72	0.78	0.81	0.81	0.79	0.74	0.68	0.59	0.50
R_4	0.50	0.78	0.96	1.06	1.10	1.09	1.02	0.92	0.80	0.66	0.50
R_5	0.50	1.02	1.34	1.51	1.55	1.50	1.37	1.18	0.97	0.74	0.50
R_6	0.50	1.42	1.97	2.23	2.27	2.13	1.89	1.57	1.21	0.85	0.50
R_7	0.50	2.12	3.04	3.43	3.43	3.14	2.70	2.16	1.59	1.02	0.50

Whence it is seen that the *greatest possible* values of R, under the assumed conditions, are—

$$\left.\begin{array}{cccccccc} R_2 & R_3 & R_4 & R_5 & R_6 & R_7 & \cdots \\ 0.6 & 0.8 & 1.1 & 1.6 & 2.3 & 3.4 & \cdots \end{array}\right\} \qquad (83)$$

While it is obvious that the combination of accidental errors f, shown in Table (B′), is very improbable, yet approximations to such combination will occur occasionally in practice. In such cases the errors (R) in functions interpolated by NEWTON's Formula may be a considerable part of the values given by (83). These values show that when the differences beyond Δ^v are neglected, the error R cannot be greater than 1.6, in units of the last place in F. In all probability this error will not exceed *one unit*; and when it is considered that the results of an average logarithmic computation are uncertain within this amount, we are justified in neglecting the error R, provided that fifth differences are practically constant.

Beyond R_5, the limiting values of R increase rapidly. We therefore conclude that, aside from the inconvenience involved, it is impracticable to interpolate by NEWTON's Formula when the differences beyond Δ^v are too large to be neglected.*

We now consider the expression Q of (77*a*), that is—

$$Q \equiv \overline{F}_0 + Aa_0 + Bb_0 + \ldots + Ll_0 \qquad (84)$$

Now, because the differences of F in Table (C) become approximately constant at $\Delta^{(i)}$, notwithstanding the irregularities they contain; so, *a fortiori*, must the differences of $\overline{F}$ in Table (A) become sensibly constant at $\Delta^{(i)}$, the quantities of this table being *mathematically exact.* Hence the differences $\Delta^{(i+1)}$ in Table (A), namely,

$$m_0, \; m_1, \; m_2, \; m_3, \; \ldots$$

will form a series of *continuous*, but very small terms, whose values are nearly equal to each other. *Per contra*, we have assumed that the differences

$$m_0 + \mu_0, \quad m_1 + \mu_1, \quad m_2 + \mu_2, \quad \ldots$$

* Excepting the case where $F(T)$ is a rational integral function of T, whose tabular values are mathematically exact.

of Table (C) either are alternately + and —, or that + and — terms succeed each other irregularly. It follows that the quantities m must be numerically *less* than the *maximum* value of μ in the series

$$\mu_0,\ \mu_1,\ \mu_2,\ \mu_3,$$

For, otherwise, if the quantities m exceeded the greatest of the quantities μ, the former would mask the effect of the latter in the combined series $m+\mu$; hence there would be no general alternation of signs in the series

$$m_0+\mu_0,\quad m_1+\mu_1,\quad m_2+\mu_2$$

But this is contrary to our assumption that the differencing in Table (C) has been carried to an order $\Delta^{(i+1)}$ which *does* exhibit a general alternation of signs. We therefore conclude that m_0 is numerically less than the maximum value of μ.

Now, from Table (B′), we observe that under the conditions assumed,

The maximum	value	of		$\alpha(=\Delta^{\prime})$	is	1	$=$	$(2)^0$;
"	"	"	"	$\beta(=\Delta^{\prime\prime})$	"	2	$=$	$(2)^1$;
"	"	"	"	$\gamma(=\Delta^{\prime\prime\prime})$	"	4	$=$	$(2)^2$;
. . . .	. . . .	. . . .	. . . .	. . . .	. . . .	. . . .	. . . .	. . . .
"	"	"	"	$\mu(=\Delta^{(i+1)})$			$=$	$(2)^i$.

Hence, m_0 is numerically less than 2^i.

We have observed above that, as a consequence of the conditions herein assumed, the differences of $\bar{F}$ in Table (A) are converging, being practically insensible beyond $\Delta^{(i)}$; hence the fundamental expansions (0), and all relations deduced from these, are valid in this case. The formula (59) is therefore applicable to the series $\bar{F}(T)$; hence, writing n for m in (59), we have

$$\delta_0' = Aa_0 + Bb_0 + Cc_0 + \ .\ .\ .\ .\ + Ll_0 + Mm_0 + Nn_0 + \ .\ .\ .\ .$$

in which as many terms should be retained as accuracy requires.

But we also have*

$$\delta_0' = \bar{F}(t+n\omega) - \bar{F}(t) = \bar{F}_n - \bar{F}_0$$

and therefore

$$\bar{F}_n = \bar{F}_0 + Aa_0 + Bb_0 + Cc_0 + \ .\ .\ .\ .\ + Ll_0 + Mm_0 + Nn_0 + \ .\ .\ .\ .$$

* See §26, where the same relations were similarly employed.

Now, by (84), this equation may be written

$$\overline{F}_n = Q + Mm_0 + Nn_0 + \ldots.$$

or

$$\overline{F}_n - Q = Mm_0 + Nn_0 + \ldots. \qquad (85)$$

The series $Mm_0 + Nn_0 + \ldots.$ therefore expresses the difference between the true mathematical value of the interpolated function and its approximate value Q. But since, as above observed, the differences m are nearly constant, it follows that the differences n are small in comparison. Hence, Nn_0 is small as compared with Mm_0; in brief, Mm_0 represents, very nearly, the value of the rapidly converging series $Mm_0 + Nn_0 + \ldots.$ in the right-hand member of (85). The latter equation may therefore be written, without sensible error,

$$\overline{F}_n - Q = Mm_0 \qquad (86)$$

From (82) we derive

$$\left.\begin{array}{l} R_3 - R_2 = +\,3C = (2^2-1)(+C) \\ R_4 - R_3 = -\,7D = (2^3-1)(-D) \\ R_5 - R_4 = +15E = (2^4-1)(+E) \\ \ldots\ldots\ldots\ldots\ldots\ldots \\ R_{i+1} - R_i = (2^i-1)(-1)^i M \end{array}\right\} \qquad (87)$$

From the last of these, we obtain

$$\pm 2^i M = R_{i+1} - R_i \pm M \qquad (88)$$

We have shown above that m_0 is numerically less than 2^i; this condition may be expressed in the form

$$m_0 = 2^i \sin\theta$$

where θ may have *any* value between 0 and 2π. From this relation we obtain

$$Mm_0 = 2^i M \sin\theta$$

or, by (88),

$$Mm_0 = (R_{i+1} - R_i \pm M)\sin\theta \qquad (89)$$

Substituting this value of Mm_0 in (86), we get

$$\overline{F}_n - Q = (R_{i+1} - R_i \pm M)\sin\theta \qquad (90)$$

From (77*a*), we have*

$$F_n - Q = R_i \tag{91}$$

which, subtracted from (90), gives

$$\bar{F}_n - F_n = R_{i+1} \sin\theta - (1 + \sin\theta) R_i \pm M \sin\theta$$

From Table (D) above we see that beyond Δ''' the coefficient M cannot exceed 0.04, which is an inappreciable quantity in the present discussion; we therefore write the last equation

$$\bar{F}_n - F_n = R_{i+1} \sin\theta - (1 + \sin\theta) R_i \tag{92}$$

The quantity R_{i+1} is numerically greater than R_i, and both are alike in sign; this condition may be expressed by the relation

$$R_i = R_{i+1} \sin^2\psi$$

in which ψ has a definite value depending upon the value of i. Substituting this expression for R_i in (92), the latter becomes

$$\bar{F}_n - F_n = R_{i+1} [\sin\theta - \sin^2\psi (1 + \sin\theta)]$$

or

$$\bar{F}_n - F_n = R_{i+1} (\sin\theta \cos^2\psi - \sin^2\psi) \tag{93}$$

Since $\cos^2\psi$ is necessarily positive, and $-\sin^2\psi$ negative, it follows that the coefficient of R_{i+1} in (93) will be numerically a maximum when $\sin\theta$ attains its greatest negative value; that is, when $\theta = \frac{3}{2}\pi$. Taking $\theta = \frac{3}{2}\pi$ in (93), we have

$$\bar{F}_n - F_n = R_{i+1} (-\cos^2\psi - \sin^2\psi) = -R_{i+1} \tag{94}$$

which is the maximum numerical value possible to $\bar{F}_n - F_n$, all conditions favoring.

$\bar{F}_n$ is the *true mathematical value* of the required function. F_n is the approximate value of this quantity which is obtained by applying NEWTON's Formula to Table (C), neglecting differences beyond $\Delta^{(i)}$: it being assumed, (1) that the given functions F_0, F_1, F_2, F_3, ... are true to the *nearest* unit of their last digit; (2) that n is positive

* The quantity R defined in (77*a*) is not distinguished by a subscript in the earlier part of this discussion. Considered as a particular term of the series R_2, R_3, R_4, ... , however, it is evident that R should be designated as R_i.

and less than unity; (3) that the differences $\Delta^{(i)}$ are approximately constant; and (4) that the differences $\Delta^{(i+1)}$ are quite small, with + and — signs following irregularly. Under these conditions, it follows from (94) that the computed value F_n can never differ from the true value $\overline{F}_n$ by more than the quantity R_{i+1}.

One point further, however, must be considered. In computing F_n by (76), we should, in practice, obtain the values of the several terms to one or two decimals further than are given in F, to avoid accumulation of errors in the final addition. But in writing the sum, F_n, the extra decimals are dropped, the result being taken to the *nearest unit*, as in F. Thus we actually use, not the quantity F_n obtained rigorously by (76), but a close approximation to that value, which we may denote by (F_n). Accordingly, the relation

$$F_n - (F_n) = \pm 0.5$$

expresses the maximum discrepancy between F_n and (F_n). Combining this expression with (94), we finally obtain

$$\overline{F}_n - (F_n) = -R_{i+1} \pm 0.5 \qquad (95)$$

The quantity $R_{i+1} \pm 0.5$ therefore represents the final limit of error in the value of an interpolated function, in units of the last decimal of F. From the value of R_6 given in (83), we find that when Δ^{v} is nearly constant, the limiting error is ± 2.8 units. Since it is highly improbable that all the necessary conditions will conspire to produce this *maximum* error, we may add that when the differences practically terminate at the fifth order, interpolated functions will *occasionally* be in error by one unit, only *rarely* in error by two units, and *never* by three.

With sixth, seventh, or higher differences employed, the results become subject to errors which in most cases would be intolerable, and which would probably be obviated by a direct calculation of the function.

From the foregoing investigation it therefore appears that, for purposes of interpolation, tabular functions should always be given with an interval sufficiently small that differences beyond Δ^{v} may be neglected. This condition is generally fulfilled in practice. As already stated in §24, the longitude and latitude of the moon are given in the *Nautical Almanac* for every twelve hours; from the values thus given, intermediate positions can always be safely interpolated by using differences no higher than the fourth or fifth order. On the other hand, a table of the moon's longitude for every 24 hours would yield differences of the eighth or even ninth order; the use of which in NEWTON's Formula might produce an error of several units in an interpolated position.

In all that follows, we shall assume that differences beyond the fifth order may be neglected. This assumption made, it follows from the preceding investigation that the fundamental formulae, (73) and (75), may be applied in all cases without sensible error, provided that n is taken less than unity.

31. We shall now solve an example which illustrates the main points of the foregoing discussion. If we tabulate the function

$$\bar{F}(T) \equiv \tfrac{1}{120000}\left\{\begin{array}{l} 606607.920 \quad - 199841.772\,T + 50804.968\,T^2 \\ + \;\; 5645.715\,T^3 - \;\; 2169.395\,T^4 + \;\; 116.817\,T^5 + 1.507\,T^6 \end{array}\right\} \quad (96)$$

for $T = 0, 1, 2, 3, \ldots . 9$, we find that the true mathematical values terminate in the fifth decimal. These values of $\bar{F}(T)$ are given in the table below, with their differences:

T	$\bar{F}(T)$	Δ'	Δ''	Δ'''	Δ^{iv}	Δ^{v}	Δ^{vi}
0	8.42511						
		−2.02003					
1	6.40508		+1.50987				
		−0.51016		−0.35955			
2	5.89492		1.15032		−0.30109		
		+0.64016		0.66064		+.23237	
3	6.53508		+0.48968		−0.06872		+.01507
		1.12984		0.72936		.24744	
4	7.66492		−0.23968		+0.17872		.01507
		0.89016		0.55064		.26251	
5	8.55508		0.79032		0.44123		.01507
		+0.09984		−0.10941		.27758	
6	8.65492		0.89973		0.71881		+.01507
		−0.79989		+0.60940		+.29265	
7	7.85503		−0.29033		+1.01146		
		−1.09022		+1.62086			
8	6.76481		+1.33053				
		+0.24031					
9	7.00512						

(A′)

This table corresponds to Table (A) of the last section. It will be observed that the values of $\bar{F}$ are peculiar from the fact that the last three decimals of each differ *only slightly* from the quantity 0.00500, or half a unit in the second decimal place; and, moreover, that the actual difference is, excepting the first function, *alternately in excess and defect.* This condition will rarely obtain, and is here selected only to illustrate the limiting case.

If now we drop the last three decimals of $\bar{F}$, we obtain a series of approximate values, denoted by F. The following table gives F, true to the nearest unit of the second decimal, together with its differences:

T	$F(T)$	Δ'	Δ''	Δ'''	Δ^{iv}	Δ^{v}	Δ^{vi}
0	8.43						
		−2.02					
1	6.41		+1.50				
		−0.52		−0.33			
2	5.89		1.17		−0.37		
		+0.65		0.70		+0.38	
3	6.54		+0.47		+0.01		−0.29
		1.12		0.69		0.09	
4	7.66		−0.22		0.10		+0.33
		0.90		0.59		0.42	
5	8.56		0.81		0.52		−0.30
		+0.09		−0.07		0.12	
6	8.65		0.88		0.64		+0.33
		−0.79		+0.57		+0.45	
7	7.86		−0.31		+1.09		
		−1.10		+1.66			
8	6.76		+1.35				
		+0.25					
9	7.01						

(C′)

Table (C′) corresponds to Table (C) of §30. It will be observed that Δ^{v} and Δ^{vi}, in (C′), represent $\Delta^{(i)}$ and $\Delta^{(i+1)}$, of Table (C). The differencing in (C′) is not carried beyond Δ^{vi}, because of the alternation of + and — terms.

The above values of F may be written as follows:

$$
\begin{aligned}
F &= \bar{F} + f \\
8.43 &= 8.42511 + 0.00489 \\
6.41 &= 6.40508 + 0.00492 \\
5.89 &= 5.89492 - 0.00492 \\
&\quad \cdots\cdots\cdots
\end{aligned}
$$

The quantities in the last column therefore represent the residual terms denoted by f in the preceding section. Expressing these values in units of the second decimal, we have the following table of f and its differences:

T	f	Δ'	Δ''	Δ'''	Δ^{iv}	Δ^{v}	Δ^{vi}
0	+0.489						
		+0.003					
1	+0.492		−0.987				
		−0.984		+2.955			
2	−0.492		+1.968		−6.891		
		+0.984		−3.936		+14.763	
3	+0.492		−1.968		+7.872		−30.507
		−0.984		+3.936		−15.744	
4	−0.492		+1.968		−7.872		+31.493
		+0.984		−3.936		+15.749	
5	+0.492		−1.968		+7.877		−31.507
		−0.984		+3.941		−15.758	
6	−0.492		+1.973		−7.881		+31.493
		+0.989		−3.940		+15.735	
7	+0.497		−1.967		+7.854		
		−0.978		+3.914			
8	−0.481		+1.947				
		+0.969					
9	+0.488						

(B″)

It will be observed that the quantities of Table (B″) are close approximations to the (limiting) values given in Table (B′), of §30.

Let us now apply NEWTON's Formula to interpolate the value of F which corresponds to $T = 0.40$, in Table (C′). Neglecting differences beyond Δ^{v}, we take from Table I (for $n = 0.40$), and from Table (C′), the quantities to be employed. The result is as follows:

				F_0 =	+8.43
A =	+0.40	a =	−2.02	Aa =	−0.8080
B =	−0.12	b =	+1.50	Bb =	−0.1800
C =	+0.064	c =	−0.33	Cc =	−0.0211
D =	−0.0416	d =	−0.37	Dd =	+0.0154
E =	+0.02995	e =	+0.38	Ee =	+0.0114
				$\therefore F_n$ =	+7.4477

Whence, we write for the value of the interpolated function,

$$\left.\begin{aligned}(F_n) &= 7.45\\ &= 7.44{,}77 + 0.00{,}23 = F_n + 0.00{,}23\end{aligned}\right\} \qquad (97)$$

Computing the true value $\overline{F}_n$ from (96), we obtain

$$\overline{F}_n = 7.4320416 + \qquad (98)$$

Hence the value $(F_n) = 7.45$, interpolated from Table (C′), is in error by 1.8 units of its last place.

The value of Q is the result obtained by interpolating $\overline{F}_n$ from Table (A′), neglecting differences after Δ^{v}. Thus we determine Q as follows:

		$\bar{F}_0 = +8.425110$
$A = +0.40$	$a_0 = -2.02003$	$Aa_0 = -0.808012$
$B = -0.12$	$b_0 = +1.50987$	$Bb_0 = -0.181184+$
$C = +0.064$	$c_0 = -0.35955$	$Cc_0 = -0.023011+$
$D = -0.0416$	$d_0 = -0.30109$	$Dd_0 = +0.012525+$
$E = +0.02995$	$e_0 = +0.23237$	$Ee_0 = +0.006959+$
		$\therefore\ Q = +7.432387+$

The value of R_5 is computed from Table (B″) in the same manner that Q has just been obtained from (A′). Thus we find

		$f_0 = +0.489$
$A = +0.40$	$\alpha_0 = +\ 0.003$	$A\alpha_0 = +0.001$
$B = -0.12$	$\beta_0 = -\ 0.987$	$B\beta_0 = +0.118$
$C = +0.064$	$\gamma_0 = +\ 2.955$	$C\gamma_0 = +0.189$
$D = -0.0416$	$\delta_0 = -\ 6.891$	$D\delta_0 = +0.287$
$E = +0.02995$	$\epsilon_0 = +14.763$	$E\epsilon_0 = +0.442$
$\therefore$ (In units of the second decimal)		$R_5 = +1.526$ [Cf. (83)]

Now, from (91) we have

$$F_n = Q + R_5 \tag{99}$$

Substituting the above values of Q and R_5, we find

$$F_n = 7.4324 + 0.01{,}53 = 7.4477$$

which agrees with the result obtained directly from Table (C′).

Since the sixth differences in Table (A′) are constant, it follows that the true value $\bar{F}_n$ differs from the above value of Q only by the term in Δ^{vi} of NEWTON's Formula. Now, the coefficient of Δ^{vi} is found from Table (D) of the last section to be approximately —0.0230. Hence, with $\Delta^{vi} = +0.01507$, we derive

$$\left.\begin{aligned}\bar{F}_n &= Q - (0.0230 \times 0.01507) \\ &= Q - \ 0.000346 \\ &= 7.432387 - 0.000346 \\ &= 7.432041\end{aligned}\right\}\ \text{(nearly)}$$

which agrees with (98). The second of these equations gives

$$Q = \bar{F}_n + 0.000346 \pm$$

Substituting this value of Q in (99), we have

$$F_n = \bar{F}_n + R_5 + 0.0346$$

where the numerical term is now expressed in the same unit as R_5. With the above determined value of $R_5 (= +1.526)$, the last equation becomes

$$F_n = \bar{F}_n + 1.56$$

Finally, since we were obliged to write (F_n) greater than F_n by 0.23 units, it follows that the actual error of interpolation in this instance is $1.56 + 0.23$, or approximately 1.8 units in the second decimal place; which agrees with the result previously obtained.

32. As a more practical application of NEWTON's Formula, we take the following

EXAMPLE. — From the appended table, find the sun's right-ascension for April $20^d\ 0^h$.

Date 1898	Sun's R.A.	Δ'	Δ''	Δ'''	Δ^{iv}
	h m s	m s	s	s	s
April 1	0 43 20.30				
		+18 13.77			
6	1 1 34.07		+ 5.15		
		18 18.92		+2.53	
11	1 19 52.99		7.68		−0.56
		18 26.60		1.97	
16	1 38 19.59		9.65		0.63
		18 36.25		1.34	
21	1 56 55.84		10.99		−0.29
		18 47.24		1.05	
26	2 15 43.08		12.04		+0.01
		18 59.28		+1.06	
May 1	2 34 42.36		+13.10		
		+19 12.38			
6	2 53 54.74				

Letting $t =$ April 16, we have

$$n = \frac{20-16}{5} = 0.80$$

Then, from Table I, and the above differences, we find

				F_0 =	1^h 38^m 19^s.59
$A = +0.80$		$a_0 = +18^m\ 36^s.25$		Aa_0 =	+0 14 53.000
$B = -0.08$		$b_0 = +\ 10.99$		Bb_0 =	− 0.879
$C = +0.032$		$c_0 = +\ 1.05$		Cc_0 =	+ 0.034
$D = -0.0176$		$d_0 = +\ 0.01$		Dd_0 =	0.000
		∴ Sun's R.A., April $20^d\ 0^h$		=	1 53 11.75

which is the value given in the *American Ephemeris* for 1898.

33. Since the value of n in the preceding example is only 0.2 less than unity, it is more convenient to interpolate *backwards* from

April 21, by means of (75). Thus, from Table I (for $n = 0.20$), and the tabular differences, we find

		F_0	$=\ 1^h\ 56^m\ 55^s.84$
$A = +0.20$	$a_{-1} = +18^m\ 36^s.25$	$-Aa_{-1}$	$= -0\ 3\ 43.250$
$B = -0.08$	$b_{-2} = +\ 9.65$	$+Bb_{-2}$	$= -\ 0.772$
$C = +0.048$	$c_{-3} = +\ 1.97$	$-Cc_{-3}$	$= -\ 0.095$
$D = -0.0336$	$d_{-4} = -\ 0.56$	$+Dd_{-4}$	$= +\ 0.019$
	∴ Sun's R.A., April $20^d\ 0^h$		$=\ 1\ 53\ 11.74$

which agrees within $0^s.01$ of the first result. Whenever a check is considered necessary, the interpolation may be performed by both methods.

Transformations of Newton's Formula.

34. *Modification of the Foregoing Notation of Differences:* Stirling's *Formula.* — In Newton's Formula of interpolation we use differences which depend only upon the functions $F_0, F_1, F_2, \ldots$; the functions preceding F_0, whether given or not, are in no way involved. We shall now transform Newton's Formula in such a manner as to involve differences both preceding and following the function from which we set out. The resulting formulae will in general be more convenient, rapidly convergent, and accurate than Newton's Formula.

In the schedule below, the preceding notation of differences is modified: the *even* differences which fall on the horizontal line through F_0 are now denoted by the subscript *zero*, as b_0 and d_0; all differences *above* this line are indicated by *accents*, as a', b', c'' etc.; while all differences *below* the horizontal line through F_0 are indicated by subscripts, as a_1, b_1, c_2, etc. The new schedule of differences will then be as follows:

T	$F(T)$	Δ'	Δ''	Δ'''	Δ^{iv}	Δ^{v}
$t-2\omega$	F_{-2}					
		a''		c''		e''
$t-\omega$	F_{-1}		b'		d'	
		a'		c'		e'
t	F_0		b_0		d_0	
		a_1		c_1		e_1
$t+\omega$	F_1		b_1		d_1	
		a_2		c_2		e_2
$t+2\omega$	F_2		b_2		d_2	
		a_3		c_3		e_3
$t+3\omega$	F_3					

To derive STIRLING'S *Formula*: Applying NEWTON'S Formula to the above schedule, we find for the value of F_n,

$$F_n = F_0 + na_1 + Bb_1 + Cc_2 + Dd_2 + Ee_3 + \ldots . \tag{100}$$

where, as before, $B, C, D, E, \ldots .$ represent the binomial coefficients of $\Delta'', \Delta''', \Delta^{iv}, \Delta^{v}, \ldots .$, respectively. Let us now put

$$a = \tfrac{1}{2}(a'+a_1) \quad , \quad c = \tfrac{1}{2}(c'+c_1) \quad , \quad e = \tfrac{1}{2}(e'+e_1) \tag{101}$$

from which, with the relations

$$a_1 - a' = b_0 \quad , \quad c_1 - c' = d_0 \quad , \quad e_1 = e' + \ldots$$

we obtain

$$a_1 = a + \tfrac{1}{2}b_0 \quad , \quad c' = c - \tfrac{1}{2}d_0 \quad , \quad c_1 = c + \tfrac{1}{2}d_0 \quad , \quad e_1 = e + \ldots \tag{102}$$

Using the equations (102), together with the relations given in §23, we find

$$\left.\begin{aligned} a_1 &= a + \tfrac{1}{2}b_0 \\ b_1 &= b_0 + c_1 = b_0 + c + \tfrac{1}{2}d_0 \\ c_2 &= c' + 2d_0 + e_1 = c + \tfrac{3}{2}d_0 + e + \ldots \\ d_2 &= d_0 + 2e_1 + \ldots = d_0 + 2e + \ldots \\ e_3 &= e_1 + \ldots = e + \ldots \end{aligned}\right\} \tag{103}$$

Upon substituting these values of $a_1, b_1, c_2, \ldots .$ in (100), the latter becomes

$$\begin{aligned} F_n &= F_0 + n(a+\tfrac{1}{2}b_0) + B(b_0+c+\tfrac{1}{2}d_0) + C(c+\tfrac{3}{2}d_0+e+\ldots) + D(d_0+2e+\ldots) + Ee + \ldots \\ &= F_0 + na + (B+\tfrac{n}{2})b_0 + (C+B)c + (D+\tfrac{3}{2}C+\tfrac{1}{2}B)d_0 + (E+2D+C)e + \ldots . \end{aligned}$$

Substituting in the last equation the values of B, C, D, E, namely,

$$B = \frac{n(n-1)}{\underline{|2}} \quad , \quad D = \frac{n(n-1)\ldots(n-3)}{\underline{|4}}$$

$$C = \frac{n(n-1)(n-2)}{\underline{|3}} \quad , \quad E = \frac{n(n-1)\ldots(n-4)}{\underline{|5}}$$

we finally obtain

$$F_n = F_0 + na + \frac{n^2}{2}b_0 + \frac{n(n^2-1)}{6}c + \frac{n^2(n^2-1)}{24}d_0 + \frac{n(n^2-1)(n^2-4)}{120}e + \ldots \tag{104}$$

which is known as STIRLING'S Formula. The even differences employed in this formula are those falling on the horizontal line through

F_0; the odd differences are the *means* of those which fall immediately above and below this line, as defined by (101).

Table II gives the values of STIRLING'S coefficients for the argument n. A glance at this table shows how much more rapidly these coefficients converge than those of NEWTON'S Formula.

EXAMPLE.—From the table below, find the R.A. of the sun for April 20d 0h.

Date 1898	Sun's R.A.	Δ'	Δ''	Δ'''	Δ^{iv}
	h m s	m s	s	s	s
April 1	0 43 20.30				
		+18 13.77			
6	1 1 34.07		+ 5.15		
		18 18.92		+2.53	
11	1 19 52.99		7.68		−0.56
		18 26.60		1.97	
16	1 38 19.59		9.65		0.63
		18 36.25		1.34	
21	1 56 55.84		10.99		−0.29
		18 47.24		1.05	
26	2 15 43.08		12.04		+0.01
		18 59.28		+1.06	
May 1	2 34 42.36		+13.10		
		+19 12.38			
6	2 53 54.74				

Taking $t =$ April 16 (as in §32), we have

$$n = \frac{20-16}{5} = 0.80$$

The horizontal lines drawn in the body of the table indicate the differences to be employed in (104), as follows:

(1) The required values of F_0, Δ'', and Δ^{iv} are those *included between* two lines;

(2) The required values of Δ' and Δ''' are the *means* of the quantities *separated by* a single line.

As before, we shall denote the coefficients of Δ', Δ'', Δ''', ... by A, B, C, ... Taking their values from Table II, with $n = 0.80$, and forming the required differences as indicated, we obtain

			F_0	= 1h 38m 19s.59
$A = +0.80$		$a = +18^m\ 31^s.425$	Aa	= + 14 49.140
$B = +0.32$		$b_0 = +\ 9.65$	Bb_0	= + 3.088
$C = -0.048$		$c = +\ 1.66$	Cc	= − 0.080
$D = -0.0096$		$d_0 = -\ 0.63$	Dd_0	= + 0.006
		∴ Sun's R.A., April 20d 0h		= 1 53 11.74

which agrees exactly with the result found in §33.

35. *Backward Interpolation by* STIRLING'S *Formula.*—When the *forward* interval approaches unity, it will be more convenient to proceed *backwards* from the following function by the formula

$$F_{-n} = F_0 - na + \frac{n^2}{2} b_0 - \frac{n(n^2-1)}{6} c + \frac{n^2(n^2-1)}{24} d_0 - \frac{n(n^2-1)(n^2-4)}{120} e + \ldots \qquad (105)$$

the coefficients of which are taken from Table II with the argument n, as before. It will be observed that (105) is derived from (104) by merely writing $-n$ for n in the latter; or, by supposing the given series to be inverted, and hence (Theorem III) changing the signs of a, c, and e.

EXAMPLE.—Solve the preceding example by (105); that is, find the sun's R.A. for April 20^d 0^h by backward interpolation.

Taking $t =$ April 21, we have

$$n = \frac{21-20}{5} = 0.20$$

The differences are formed for the date April 21 in the same manner as found above for April 20; thence, taking the coefficients from Table II, with $n = 0.20$, we find

			m s	F_0	=	h m s 1 56 55.84
A =	+0.20	a =	+18 41.745	$-Aa$	= −	3 44.349
B =	+0.02	b_0 =	+ 10.99	$+Bb_0$	= +	0.220
C =	−0.032	c =	+ 1.20	$-Cc$	= +	0.038
D =	−0.0016	d_0 =	− 0.29	$+Dd_0$	=	0.000
		∴ Sun's R.A., April 20^d 0^h			=	1 53 11.75

36. EXAMPLE.—Use STIRLING'S Formula to compute log sin 9° 22′ from the following table:

T	Log sin T	Δ'	Δ''	Δ'''	Δ^{iv}	Δ^{v}
6°	9.01923					
		+6666				
7	9.08589		−899			
		5767		+209		
8	9.14356		690		−62	
		5077		147		+17
9	9.19433		543		45	
		4534		102		+19
10	9.23967		441		−26	
		4093		+ 76		
11	9.28060		−365			
		+3728				
12	9.31788					

Here we have

$$t = 9° \qquad n = \frac{22}{60} = 0.36667$$

and we therefore obtain

				F_0	=		9.19433
A =	+0.36667	a =	+4805.5	Aa	=	+	1762.0
B =	+0.06722	b_0 =	− 543	Bb_0	=	−	36.5
C =	−0.05289	c =	+ 124.5	Cc	=	−	6.6
D =	−0.00485	d_0 =	− 45	Dd_0	=	+	0.2
E =	+0.01022	e =	+ 18	Ee	=	+	0.2
			∴ Log sin 9° 22′		=		9.21152.3

The true value to six decimals is 9.211526.

37. *The Algebraic Mean.*—It may be well to observe that in taking the mean of two quantities having like signs, and of nearly the same magnitude, it is easier to add one-half their *difference* to the lesser number, than to take one-half the sum of the two quantities. That is, we proceed according to the identity

$$\tfrac{1}{2}(x+y) = x + \tfrac{1}{2}(y-x)$$

in which we suppose y numerically greater than x. Thus, in the last example, instead of taking

$$a = \tfrac{1}{2}(a'+a_1) = \tfrac{1}{2}(5077+4534) = \tfrac{1}{2}(9611) = +4805.5$$

it is easier to follow the equivalent formula

$$a = a_1 - \tfrac{1}{2}(a_1-a') = a_1 - \tfrac{1}{2}b_0 = 4534 + \tfrac{1}{2}(543) = +4805.5$$

Similarly, we find

$$c = 102 + 22.5 = +124.5$$

Per contra, to form the mean of two quantities having unlike signs, and differing but little in magnitude, it is easier to take their algebraic sum and then divide by two. For example, given the values

$F(T)$	Δ'	Δ''
F_{-1}		
F_0	−4226	+9314
F_1	+5088	

we find

$$a = \tfrac{1}{2}(5088-4226) = \tfrac{1}{2}(+862) = +431$$

With these precepts, the required *mean differences* of interpolation are very readily taken.

38. BESSEL'S *Formula*.—We now pass from STIRLING'S Formula to another, somewhat similar, wherein we employ the odd differences a_1, c_1, e_1, which fall on the horizontal line between F_0 and F_1, and the *means* of the even differences falling immediately above and below this line. Using the schedule on page 62, let us put

$$b = \tfrac{1}{2}(b_0+b_1) \quad , \quad d = \tfrac{1}{2}(d_0+d_1) \tag{106}$$

Then, since $b_1-b_0=c_1$, and $d_1-d_0=e_1$, these equations give

$$b_0 = b - \tfrac{1}{2}c_1 \quad , \quad d_0 = d - \tfrac{1}{2}e_1 \tag{107}$$

Let us write the formula (104), for brevity,

$$F_n = F_0 + na + \tfrac{n^2}{2}b_0 + Cc + Dd_0 + Ee + \ldots \tag{108}$$

where

$$C = \frac{n(n^2-1)}{6} \quad , \quad D = \frac{n^2(n^2-1)}{24} \quad , \quad E = \frac{n(n^2-1)(n^2-4)}{120} \tag{109}$$

Now, by means of (102) and (107), we derive

$$\left.\begin{aligned} a &= a_1 - \tfrac{1}{2}b_0 = a_1 - \tfrac{1}{2}(b-\tfrac{1}{2}c_1) = a_1 - \tfrac{1}{2}b + \tfrac{1}{4}c_1 \\ b_0 &= b - \tfrac{1}{2}c_1 \\ c &= c_1 - \tfrac{1}{2}d_0 = c_1 - \tfrac{1}{2}(d-\tfrac{1}{2}e_1) = c_1 - \tfrac{1}{2}d + \tfrac{1}{4}e_1 \\ d_0 &= d - \tfrac{1}{2}e_1 \\ e &= e_1 - \ldots \end{aligned}\right\} \tag{110}$$

Upon substituting these values of a, b_0, c, ... in (108), we have

$$\begin{aligned} F_n &= F_0 + n(a_1-\tfrac{1}{2}b+\tfrac{1}{4}c_1) + \tfrac{n^2}{2}(b-\tfrac{1}{2}c_1) + C(c_1-\tfrac{1}{2}d+\tfrac{1}{4}e_1) + D(d-\tfrac{1}{2}e_1) + E(e_1-\ldots) + \ldots \\ &= F_0 + na_1 + (\tfrac{n^2}{2}-\tfrac{n}{2})b + (C-\tfrac{n^2}{4}+\tfrac{n}{4})c_1 + (D-\tfrac{1}{2}C)d + (E-\tfrac{1}{2}D+\tfrac{1}{4}C)e_1 + \ldots \end{aligned}$$

Finally, substituting in the last equation the values of C, D, E, from (109), we obtain

$$\begin{aligned} F_n = F_0 + na_1 &+ \frac{n(n-1)}{2}b + \frac{n(n-1)(n-\frac{1}{2})}{6}c_1 \\ &+ \frac{(n+1)n(n-1)(n-2)}{24}d + \frac{(n+1)n(n-1)(n-2)(n-\frac{1}{2})}{120}e_1 + \ldots \end{aligned} \tag{111}$$

which is BESSEL'S Formula of interpolation, commonly regarded as the most convenient and accurate of the several forms in use. The odd differences here employed are those which fall on the horizontal line between F_0 and F_1, as shown in the schedule on page 62; the even differences are the *means* of those falling immediately above and below this line, as defined by (106).

Table III gives BESSEL'S coefficients for the argument n.

EXAMPLE.—Use BESSEL'S Formula to compute log sin 9° 22′ from the table below:

T	Log sin T	Δ'	Δ''	Δ'''	Δ^{iv}	Δ^{v}
6°	9.01923					
		+6666				
7	9.08589		−899			
		5767		+209		
8	9.14356		690		−62	
		5077		147		+17
9	9.19433		543		45	
		4534		102		+19
10	9.23967		441		−26	
		4093		+ 76		
11	9.28060		−365			
		+3728				
12	9.31788					

We have, as in §36,

$$t = 9^\circ \qquad n = 0.36667$$

The horizontal lines drawn in the table indicate that the values of F_0, Δ', Δ''' and Δ^{v}, to be employed in (111), are those included between the parallel lines; while the required values of Δ'' and Δ^{iv} are the *means* of the quantities separated by the single line. Forming the differences thus indicated and taking their coefficients from Table III, with $n = 0.36\frac{2}{3}$, we obtain

				F_0 =	9.19433
A =	+0.36667	a_1 =	+4534	Aa_1 =	+ 1662.5
B =	−0.11611	b =	− 492	Bb =	+ 57.1
C =	+0.00516	c_1 =	+ 102	Cc_1 =	+ 0.5
D =	+0.02160	d =	− 36	Dd =	− 0.8
E =	−0.00057	e_1 =	+ 19	Ee_1 =	0.0
				∴ Log sin 9° 22′ =	9.21152.3

which agrees exactly with the value found in §36.

39. EXAMPLE.—Find by BESSEL'S Formula the value of 10^4 from the following table of T^4.

T	T^4	Δ'	Δ''	Δ'''	Δ^{iv}	Δ^{v}
− 8	+ 4096					
		− 4015				
− 3	81		+ 3950			
		− 65		− 1500		
+ 2	16		2450		+15000	
		+ 2385		+13500		0
7	2401		15950		15000	
		18335		28500		0
12	20736		44450		+15000	
		62785		+43500		
17	83521		+87950			
		+150735				
+22	+234256					

Taking $t = 7$, we have

$$n = \tfrac{10-7}{5} = 0.60$$

Therefore we find

						F_0	=	+ 2401
A	=	+0.60	a_1	=	+18335	Aa_1	=	+11001
B	=	−0.120	b	=	+30200	Bb	=	− 3624
C	=	−0.0040	c_1	=	+28500	Cc_1	=	− 114
D	=	+0.0224	d	=	+15000	Dd	=	+ 336
						$\therefore 10^4$	=	+10000

40. *Backward Interpolation by* BESSEL'S *Formula.* — To find F_{-n} by BESSEL'S Formula, we conceive the series given on page 62 to be inverted; the required function is then found by interpolating *foward* from F_0 toward F_{-1} with the interval n. Hence, the differences to be used in (111) are —

$$-a', \quad +\tfrac{1}{2}(b_0+b'), \quad -c', \quad +\tfrac{1}{2}(d_0+d'), \quad -e',$$

We therefore have

$$F_{-n} = F_0 - na' + \frac{n(n-1)}{2} \cdot \frac{b_0+b'}{2} - \frac{n(n-1)(n-\frac{1}{2})}{6} c' + \ldots \qquad (111a)$$

the coefficients, as in (111), being taken from Table III with the argument n.

EXAMPLE. — Find 10^4 from the table of §39, by means of (111*a*).

Taking $t = 12$, we find

$$n = \tfrac{12-10}{5} = 0.40$$

The differences are here the same as in the last example; thus we obtain

						F_0	=	+20736
A	=	+0.40	a'	=	+18335	$-Aa'$	=	− 7334
B	=	−0.120	$\frac{b_0+b'}{2}$	=	+30200	$+B.\frac{b_0+b'}{2}$	=	− 3624
C	=	+0.0040	c'	=	+28500	$-Cc'$	=	− 114
D	=	+0.0224	$\frac{d_0+d'}{2}$	=	+15000	$+D.\frac{d_0+d'}{2}$	=	+ 336
						$\therefore 10^4$	=	+10000

41. *Property of* BESSEL'S *Coefficients.* — If we take from Table III the coefficients for Δ'', Δ''', Δ^{iv}, Δ^{v}, with the argument $n = 0.30$, and also with $n = 0.70$ (= 1.00—0.30), we find the following values:

n	B	C	D	E
0.30	−.10500	+.00700	+.01934	−.00077
0.70	−.10500	−.00700	+.01934	+.00077

It will be observed that the coefficients are here numerically the same for the arguments n and $1-n$; having like signs for the even orders, and opposite signs for the odd orders of differences.

More generally, let us denote the values of BESSEL'S coefficients for Δ'', Δ''', Δ^{iv}, Δ^{v}, ... taken with the argument n, by $B, C, D, E, \ldots$, respectively; and the corresponding values taken with the argument $1-n$ by B_1, C_1, D_1, E_1, ... An inspection of Table III then shows that we have

$$\left.\begin{aligned} B_1 &= +B \\ C_1 &= -C \\ D_1 &= +D \\ E_1 &= -E \\ &\cdots\cdots \end{aligned}\right\} \tag{112}$$

To establish these relations generally, we write (111) in the form

$$F_n = F_0 + na_1 + Bb + Cc_1 + Dd + Ee_1 + \ldots . \tag{113}$$

Now, the value of F_n may also be obtained by interpolating *backwards* from F_1 with the interval $1-n$; the differences thus involved will be exactly the same as in (113). Hence, after the manner of formula (111*a*), we have

$$F_n = F_1 - (1-n)\,a_1 + B_1 b - C_1 c_1 + D_1 d - E_1 e_1 + \ldots . \tag{114}$$

But we have, also,

$$F_1 - (1-n)\,a_1 = (F_1 - a_1) + na_1 = F_0 + na_1$$

Whence, (114) becomes

$$F_n = F_0 + na_1 + B_1 b - C_1 c_1 + D_1 d - E_1 e_1 + \ldots . \tag{115}$$

which, subtracted from (113), gives

$$0 = (B-B_1)\,b + (C+C_1)\,c_1 + (D-D_1)\,d + \ldots . \tag{116}$$

The equation (116) is true in all cases to which the formulae of interpolation are applicable; it is therefore true when $F(T)$ is a rational integral function of the second degree. But, in the latter case, the second differences being constant, we have

$$c_1 = d = e_1 = \ldots . = 0$$

The equation (116) then becomes

$$0 = (B-B_1)\,b$$

Hence, since b cannot vanish, we have

$$B_1 = +B$$

This result reduces (116) to the form

$$0 = (C+C_1)c_1 + (D-D_1)d + (E+E_1)e_1 + \ .\ . \qquad (117)$$

Again, we may suppose Δ''' constant; that is, we may put

$$d = e_1 = \ .\ .\ .\ .\ = 0$$

The equation (117) then becomes

$$0 = (C+C_1)c_1$$

or

$$C_1 = -C$$

By repeated application of this reasoning, we prove that the relations (112) are true generally.

It follows that the numerical process involved in finding F_n by Bessel's Formula is identical whether we interpolate forward from F_0 or backward from F_1, except for the terms in F and Δ'. Hence little or no check is afforded by performing the interpolation by both methods. When such a check is deemed necessary, Bessel's and Stirling's Formulae should both be used.

42. *Relative Advantages of* Newton's, Stirling's, *and* Bessel's *Formulae.* — In practice, the only important application of Newton's Formula consists in interpolating functional values near the *beginning* or *end* of a given series. The selection of this formula is then a matter of necessity rather than of preference.

In all other cases, either of the more rapidly converging formulae of Stirling or Bessel should be employed. Regarding a choice between these two, when Tables II and III are available there would appear to be very little advantage one way or the other. The form given by Bessel is more commonly used, and is perhaps a trifle more accurate in practice than Stirling's form, particularly for values of n in the neighborhood of *one-half*. When n is quite small, however, Stirling's Formula will probably be found more convenient.

Suppose we have given a limited table of functions, as follows :

$F(T)$	Δ'	Δ''	Δ'''	Δ^{iv}
F_{-2}				
	a''			
F_{-1}		b'		
	a'		c'	
F_0		b_0		d_0
	a_1		c_1	
F_1		b_1		d_1
	a_2		c_2	
F_2		b_2		
	a_3			
F_3				

Assuming that fourth differences must be taken into account, and that fifth differences are to be neglected, the value of F_n should in this case be computed by BESSEL's Formula, which employs the mean of the quantities d_0 and d_1. If, however, the function F_3 were not included in this series, then the term d_1 would not be given, and we should proceed by STIRLING's Formula, which involves d_0 directly.

BESSEL's Formula is particularly simple and convenient when $n = \frac{1}{2}$, that is, when it is required to find the function which falls midway between F_0 and F_1; this important case will be fully considered in a later section.

43. *Simple Interpolation.*—When frequent interpolation is required, as in tables of logarithms, trigonometric functions, etc., the interval of the argument is usually chosen sufficiently small that the effect of second differences may be neglected. BESSEL's Formula gives in this case

$$F_n = F_0 + na_1 \tag{118}$$

To interpolate *backwards* from F_0, that is, to find F_{-n}, we obtain from (111*a*), by neglecting second and higher differences,

$$F_{-n} = F_0 - na' \tag{119}$$

Upon these formulae the process of *simple interpolation* is based. The first difference to be used in either case is the value falling between F_0 and the function toward which the interpolation proceeds.

Frequently, where great accuracy is not required, it is sufficient to obtain F_n by simple interpolation even when the second differences are considerable. In such a case, supposing that the third differences are insensible, we observe from BESSEL'S Formula that the error of the approximate value of F_n will be —

$$\delta F_n = \frac{n(n-1)}{2} \Delta'' \tag{120}$$

The maximum value of $\frac{n(n-1)}{2}$, which obtains for $n = \frac{1}{2}$, is $-\frac{1}{8}$; whence we have the following result:

When second differences are sensibly constant, the maximum error of functions obtained by simple interpolation is $\frac{1}{8}\Delta''$.

Thus, in Tables I, II, and III, the values of the coefficients for Δ'' (designated above as B) can never be in error by more than $\frac{1}{8}$ of 10 units, or 1.2 units in the fifth decimal, when found by simple interpolation.

44. *Interpolation Involving Second Differences, by Means of a Corrected First Difference.* — When the second differences are constant, or nearly so, but too large to neglect, their effect may be included (and hence an accurate value of F_n obtained) by the following simple method:

Since third differences are supposed insensible, BESSEL'S Formula becomes

$$F_n = F_0 + na_1 + \frac{n(n-1)}{2} b$$

which may be written in the form

$$F_n = F_0 + n\left[a_1 - \left(\frac{1-n}{2}\right) b\right] \tag{121}$$

Now, because third differences are negligible, we may write b_0 for b in (121); then, putting

we have

$$\left.\begin{aligned} \alpha_1 &= a_1 - \left(\frac{1-n}{2}\right) b_0 \\ F_n &= F_0 + n\alpha_1 \end{aligned}\right\} \tag{122}$$

The value of F_n is thus obtained almost as readily as in simple interpolation. In forming the quantity $\frac{1-n}{2}$ (which is simply one-half the complement of n with respect to unity), only an approximate value of n is ordinarily required. The value of α_1, the *corrected first difference*, is thus found by an easy mental process amounting almost to mere inspection.

Example.—Find $(8.2)^2$ from the following values of T^2:

T	T^2	Δ'	Δ''
4	16		
		+33	
7	49		+18
		51	
10	100		+18
		+69	
13	169		

Here we have

$$t = 7 \qquad n = 0.4 \qquad F_0 = 49 \qquad a_1 = 51 \qquad b_0 = 18$$

Hence, by (122), we find

$$\frac{1-n}{2} = \frac{1-0.4}{2} = 0.3$$

$$a_1 = 51 - (0.3 \times 18) \quad = 45.6$$

$$\therefore\ F_n = 49 + (0.4 \times 45.6) = 67.24$$

This result is exact, because the second differences are rigorously constant.

45. *Backward Interpolation by Means of a Corrected First Difference.* — From (111*a*), neglecting differences beyond Δ'', we obtain

$$F_{-n} = F_0 - na' + \frac{n(n-1)}{2} \cdot \frac{b_0 + b'}{2} = F_0 - na' + \frac{n(n-1)}{2} b_0$$

or

$$F_{-n} = F_0 - n\left(a' + \frac{1-n}{2} b_0\right) \tag{123}$$

Hence, if we put

we have

$$\left.\begin{aligned} a' &= a' + \left(\frac{1-n}{2}\right) b_0 \\ F_{-n} &= F_0 - na' \end{aligned}\right\} \tag{124}$$

Example.—From Hill's *Tables of Saturn*, the following perturbations are taken; find the value corresponding to the argument $T = 30682.38$.

T	$F(T)$	Δ'	Δ''
28800	12.5751		
		−3753	
29760	12.1998		+70
		3683	
30720	11.8315		68
		3615	
31680	11.4700		+63
		−3552	
32640	11.1148		

Taking $t = 30720$, we have

$$F_0 = 11.8315 \qquad n = \frac{720-682.38}{960} = 0.03919 \text{ (backward from } F_0)$$
$$T = 30682.38 \qquad a' = -3683$$
$$\omega = 960 \qquad b_0 = +\ \ 68$$

Using 0.04 as a sufficiently accurate value of n in determining $\mathfrak{a}'$, we find by (124),

$$\frac{1-n}{2} = \frac{1-0.04}{2} = 0.48$$
$$\mathfrak{a}' = -3683 + (0.48 \times 68) = -3650$$
$$\therefore F_{-n} = 11.8315 - [0.03919 \times (-3650)] = 11.8458$$

In the present example the algebraic signs of the several quantities of (124) have each been considered. Now it is important to remark that in the majority of cases no attention need be given to these signs; for in this fact lies the chief practical advantage of the method. Thus, in the present example, we are interpolating from the third function toward the second; the value of Δ' to be corrected is the difference of these two functions, or 3683; *the sign we disregard.* The correction to be applied to this number is 0.48×68, or 33. Again neglecting signs, we simply apply this quantity to 3683 in such a manner as to obtain a result falling somewhere *between* the numbers 3683 and 3615 of the column Δ'. Hence, we *decrease* 3683 by 33, thus obtaining 3650 for our corrected first difference, $\mathfrak{a}'$. Finally, $n\mathfrak{a}' = 143$, by which amount we *increase* the function 11.8315 (giving 11.8458), since we observe that the functions are *increasing* in the direction of the interpolation.

A partial exception to this mechanical method of procedure is to be observed when a_1 and a' have opposite signs; that is, when Δ' changes sign in passing the function F_0. In this case the sign of $\mathfrak{a}$ must be noted; we then have, as in (122) and (124),

$$\left.\begin{aligned} F_n &= F_0 + n\mathfrak{a}_1 \\ F_{-n} &= F_0 - n\mathfrak{a}' \end{aligned}\right\} \qquad (124a)$$

For example, given the values below:

T	$F(T)$	Δ'	Δ''
10	138		
		+400	
15	538		−300
		+100	
20	638		−300
		−200	
25	438		

Suppose it is required to find F, for $T = 19$. We let $t = 20$, $F_0 = 638$, and interpolate *backwards* with $n = 0.20$. To obtain a', decrease 100 by 0.4×300, or 120; whence $a' = -20$, and therefore

$$F_{-n} = F_0 - na' = 638 - [0.2 \times (-20)] = 642$$

We remark in passing that the value of the *corrected first difference*, either in forward or backward interpolation, is always contained between the limits a_1 and a'.

The number of instances in practice where the differences beyond Δ'' may be neglected is very large. The precepts given above are therefore important, and should be practiced by the student until their application becomes rapid and mechanical.

46. *Correction of Erroneous Functions by Direct Interpolation of the Values in Question.*—When an error has been detected in some one function of a series by the method of differences, as explained in §8, it is often possible to find the true value of that quantity by direct interpolation. To accomplish this, we have only to omit from the given series every *alternate* function, the incorrect value being one of the number rejected. We have then to make but one interpolation, *midway between* two functions of the new series, to obtain the value required. It is necessary, however, that the given series shall include a sufficient number of functions to furnish an adequate schedule of differences in the abridged table; furthermore, the interval of the original table must be sufficiently small that the magnified differences of the abridged table will not be so large as to render interpolation impossible.

We illustrate by means of Example III, §9. The value of β for May 11.0 was found to be incorrect; hence, to find the true value, we omit from the given series the positions for every noon, retaining only the values for each midnight. Thus we obtain the following abridged series:

Date 1898	β	Δ'	Δ''	Δ'''	Δ^{iv}
	° ′ ″	° ′ ″	′ ″	′ ″	″
May 8.5	−1 59 54.2				
		+1 15 27.2			
9.5	−0 44 27.0		+1 39.7		
		1 17 6.9		−5 14.1	
10.5	+0 32 39.9		−3 34.4		+54.8
		1 13 32.5		−4 19.3	
11.5	1 46 12.4		−7 53.7		
		+1 5 38.8			
12.5	+2 51 51.2				

The value of β for May 11.0 is now readily found by interpolation; for this purpose, we take

$$t = \text{May } 10.5 \qquad F_0 = +0°\ 32'\ 39''.9 \qquad n = 0.50$$

Since but one value of Δ^{iv} is given, namely $d_0 = 54.8$, we proceed by STIRLING's Formula (see § 42); thus we find

				F_0 =	+0° 32′ 39″.9
$A = +\frac{1}{2}$	a =	+1° 15′ 19″.7		Aa =	+0 37 39.85
$B = +\frac{1}{8}$	b_0 =	− 3 34.4		Bb_0 =	− 26.80
$C = -\frac{1}{16}$	c =	− 4 46.7		Cc =	+ 17.92
$D = -0.00781$	d_0 =	+ 54.8		Dd_0 =	− 0.43
			∴ β (May 11.0, 1898)	=	+1 10 10.44

The value found in § 9 by the method of differences is $+1°\ 10'\ 10''.6$. The result just obtained by interpolation is uncertain within narrow limits, because we have no knowledge of the value of Δ^{v} in the above table. The value $1°\ 10'\ 10''.6$ should therefore be taken as the more probable.

Had the value of β for May 13.5 been included in the original series, our abridged table would have yielded two values of Δ^{iv} and one of Δ^{v}. We should then have used BESSEL's Formula (see § 42) to compute the latitude for May 11.0. Now, the moon's latitude for May 13.5, 1898, is $+3°\ 46'\ 22''.2$; including this value with the others above, and applying BESSEL's Formula, we find $\beta = +1°\ 10'\ 10''.57$.

47. When a series contains *several* incorrect functions, separated from each other by *even* multiples of the interval ω, the foregoing

method at once serves for the determination of the several values in question. Thus, in the series

$$F_0,\ F_1,\ F_2,\ F_3,\ F_4,\ \ldots\ .$$

let us suppose that F_1, F_3, and F_7 are in error. Then, if we tabulate and difference the series

$$F_0,\ F_2,\ F_4,\ F_6,\ F_8,\ \ldots\ .$$

the required values are easily found by interpolation.

Again, when two *adjacent* functions, say F_4 and F_5, require correction, we may proceed by tabulating every *third* function of the given series; thus we obtain the abridged series

$$F_0,\ F_3,\ F_6,\ F_9,\ \ldots\ .$$

from which the values of F_4 and F_5 are found by interpolating with $n = \frac{1}{3}$ and $\frac{2}{3}$, respectively. Otherwise, if the differences of the latter series are too large for accurate interpolation, we may omit from the original table every *alternate* function only, as in §46. The resulting series,

$$F_0,\ F_2,\ F_4,\ F_6,\ F_8,\ \ldots\ .$$

will therefore contain but one incorrect value, namely F_4. The correction to F_4 may then be found by the method of differences, whereas this method might be impracticable if applied to F_4 and F_5 simultaneously. Similarly, we may correct F_5 by the differences of

$$F_1,\ F_3,\ F_5,\ F_7,\ F_9,\ \ldots\ .$$

or, by interpolation from the corrected series

$$F_0,\ F_2,\ F_4,\ F_6,\ F_8,\ \ldots\ .$$

Systematic Interpolation—Subdivision of Tables.

48. Thus far we have considered interpolation as a process for computing the values of functions for occasional or *special* values of the argument, simply. We shall now consider the subject in a broader

sense, and find that interpolation is of great importance as applied in a more extended and systematic manner.

When a complicated function is to be computed and tabulated for a large number of equidistant values of the argument, or when the tabular quantities result from a long and laborious calculation, it will be much shorter and easier to make the direct computation for a less frequent interval than is finally required, and thence to obtain the intermediate values by systematic interpolation. For example, suppose the function

$$F(T) = 700''.43 \sin 2T - 1''.19 \sin 4T$$

is to be tabulated for every 10′ from 30° to 60°; we should begin by computing $F(T)$ for every 4th degree of T. Thus we should obtain the values of $F(T)$ for $T =$

$$22°, 26°, 30°, 34°, \ldots . 70° ;$$

the calculation being extended somewhat beyond the assigned limits in order to facilitate the interpolation which follows. These quantities having been differenced, and corrected for accidental errors if necessary, the *middle terms* are then found by interpolation to *halves*. We thus obtain the series $F(T)$ corresponding to $T =$

$$26°, 28°, 30°, 32°, \ldots . 64°$$

Interpolating again to halves, we have a table of $F(T)$ for every degree of T. A third interpolation to halves gives the function for every 30′. Finally, interpolating the latter series *to thirds*, we obtain the required table, giving $F(T)$ for every 10′ of the argument T. It is obvious that the labor of computation decreases rapidly with each successive interpolation.

All of the extended tables in common use, such as tables of logarithms, sines, tangents, etc., have been subdivided in this manner, at a saving of labor almost beyond estimation. In fact, interpolation has undoubtedly done more for mathematical science than any other discovery, excepting that of logarithms.

The following sections will be devoted to the derivation of formulae and precepts which will simplify the process of systematic interpolation just described. Instead of performing a separate and distinct calculation for each interpolated function, we shall develop a method by which the required values are obtained by *successive additions* of the *computed differences* of those values.

The most convenient interpolation to perform, either in an isolated case, or as applied to the subdivision of an extended series, is interpolation *to halves*, which gives the function corresponding to the *mean* of two consecutive tabular values of the argument. This case will now be considered.

49. *Interpolation to Halves.*—If, in BESSEL'S Formula (111), we put $n = \frac{1}{2}$, the coefficients of Δ''' and Δ^{v} vanish, and we get

$$F_{\frac{1}{2}} = F_0 + \tfrac{1}{2}a_1 - \tfrac{1}{8}b + \tfrac{3}{128}d - \ldots \qquad (125)$$

Since $F_1 - F_0 = a_1$, we have

$$F_0 + \tfrac{1}{2}a_1 = \frac{F_0 + F_1}{2}$$

Also, by (106), we have

$$b = \frac{b_0 + b_1}{2}$$

$$d = \frac{d_0 + d_1}{2}$$

Hence, (125) may be written in the form

$$F_{\frac{1}{2}} = \frac{F_0 + F_1}{2} - \tfrac{1}{8}\left(\frac{b_0 + b_1}{2}\right) + \tfrac{3}{128}\left(\frac{d_0 + d_1}{2}\right) - \ldots \qquad (126)$$

which is the formula for *interpolation to halves*, true to fifth differences inclusive. The differences are to be taken according to the schedule on page 62.

Supposing that fourth differences are so small as to produce no sensible effect, we obtain from (126) the very simple formula

$$F_{\frac{1}{2}} = \frac{F_0 + F_1}{2} - \tfrac{1}{8}\left(\frac{b_0 + b_1}{2}\right) \qquad (127)$$

true to third differences inclusive. Hence, to interpolate a function *midway* between two consecutive tabular values, we have the following

RULE: *From the mean of the two given functions, subtract one-eighth the mean of the second differences which stand opposite.* The result is true to third differences inclusive. To obtain the value true to fifth differences inclusive, *add to the above result* $\frac{3}{128}$ *of the mean of the corresponding fourth differences.*

50. *Precepts for Systematic Interpolation to Halves.*—The foregoing rule applies either to the interpolation of a single function into the middle, or to that of an entire series of values. For the latter purpose, however, the work may be arranged in a more expeditious manner, as follows:

For convenience, we assume for the present that 4th differences may be neglected; accordingly, if we put

$$\delta_0' = F_{\frac{1}{2}} - F_0 \ , \quad \delta_1' = F_1 - F_{\frac{1}{2}} \ , \quad \delta_2' = F_{\frac{3}{2}} - F_1 \ , \quad \delta_3' = F_2 - F_{\frac{3}{2}}, \ \ldots \tag{128}$$

we obtain from (125),

$$\left.\begin{aligned} \delta_0' &= \tfrac{1}{2}a_1 - \tfrac{1}{8}\left(\frac{b_0+b_1}{2}\right) \\ \delta_2' &= \tfrac{1}{2}a_2 - \tfrac{1}{8}\left(\frac{b_1+b_2}{2}\right) \\ \delta_4' &= \tfrac{1}{2}a_3 - \tfrac{1}{8}\left(\frac{b_2+b_3}{2}\right) \\ &\ldots\ldots\ldots\ldots \end{aligned}\right\} \tag{129}$$

The quantities δ' defined by (128) are evidently the *first differences* of the *interpolated* series; the alternate terms, $\delta_0', \delta_2', \delta_4', \ldots$, are computed by (129) from the first and second differences of the *given* series of functions; the values of $\delta_1', \delta_3', \delta_5', \ldots$ are not computed. The method and arrangement of the work are shown in the schedule below:

T	$F(T)$	δ'	δ''	α	β	Δ'	Δ''	Δ'''
$t-\omega$	F_{-1}							
						a'		c'
t	F_0						b_0	
$t+\frac{1}{2}\omega$	$F_{\frac{1}{2}}$	δ_0'	δ_0''	$\frac{1}{2}a_1$	$-\frac{1}{8}\left(\frac{b_0+b_1}{2}\right)$	a_1		c_1
$t+\omega$	F_1	δ_1'	δ_1''				b_1	
$t+\frac{3}{2}\omega$	$F_{\frac{3}{2}}$	δ_2'	δ_2''	$\frac{1}{2}a_2$	$-\frac{1}{8}\left(\frac{b_1+b_2}{2}\right)$	a_2		c_2
$t+2\omega$	F_2	δ_3'	δ_3''				b_2	
$t+\frac{5}{2}\omega$	$F_{\frac{5}{2}}$	δ_4'	δ_4''	$\frac{1}{2}a_3$	$-\frac{1}{8}\left(\frac{b_2+b_3}{2}\right)$	a_3		c_3
$t+3\omega$	F_3	δ_5'	δ_5''				b_3	

The differences of the given series are placed in the last three columns, under Δ', Δ'', and Δ'''. The column α is then filled in by writing opposite each of the quantities Δ' one-half its value. The column β is also computed, each term being *minus* one-eighth the mean of the two values of Δ'' which stand opposite. The *alternate* quantities of column δ' are then found, as in (129), by taking the sums of the corresponding terms in α and β; the results are written immediately *above* the line of the latter terms, so as to fall between F_0 and F_1, F_1 and F_2, etc., respectively.

Finally, since by (128) we have

$$F_{\frac{1}{2}} = F_0 + \delta_0' \;,\quad F_{\frac{3}{2}} = F_1 + \delta_2' \;,\quad F_{\frac{5}{2}} = F_2 + \delta_4' \;, \tag{130}$$

it is only necessary to add each computed value of δ' to the function immediately preceding, to obtain the required middle functions. Having thus completed the interpolation, the remaining or alternate values of δ' are filled in by direct differencing. The second differences are then written in the column δ'', their regularity proving the accuracy of the work.

The *given functions*, also the *computed* first differences, etc., are distinguished in the above schedule by heavy type.

When it is necessary to take account of 4th and 5th differences, we have only to form an extra column γ, to follow β in the schedule above. Under γ we write the terms

$$\frac{3}{128}\left(\frac{d_0+d_1}{2}\right) \;,\quad \frac{3}{128}\left(\frac{d_1+d_2}{2}\right) \;,\ \text{etc.};$$

the values of δ' are then formed by adding the three corresponding terms in α, β, and γ.

Example.—Given the values of log sin T for $T = 30°, 32°, 34°, \ldots\ 42°$; find the value for every degree of T from 32° to 40°, inclusive.

In accordance with the method above outlined, we arrange the given functions, with their differences, as follows:

T	Log sin T	δ'	δ''	α	β	Δ'	Δ''	Δ'''
30°	**9.69897**							
31						+2524		
32	**9.72421**	**+1190**					−189	
33	9.73611	1145	−45	**+1167.5**	**+22.4**	2335		+20
34	**9.74756**	**1103**	42				169	
35	9.75859	1063	40	**1083.0**	**20.2**	2166		15
36	**9.76922**	**1024**	39				154	
37	9.77946	988	36	**1006.0**	**18.3**	2012		15
38	**9.78934**	**953**	35				139	
39	9.79887	+ 920	−33	**+ 936.5**	**+16.7**	1873		+10
40	**9.80807**						−129	
41						+1744		
42	**9.82551**							

Since 4th differences may be neglected, only the two columns α and β are required for the computation of the differences δ'. All the quantities actually used in the process are given in the above table. The computed quantities, together with the given values of log sin T, are printed in heavy type, to render this process more evident.

51. *To Reduce the Argument Interval of a Given Table from ω to $m\omega$, where $\frac{1}{m}$ is a Positive Odd Integer.*—As particular cases of this problem, we may take $m = \frac{1}{3}, \frac{1}{5}, \frac{1}{9}$, etc. Taking $m = \frac{1}{3}$, we introduce *two* values between every two adjacent functions of the given table; we thus derive the series

$$F_0, F_{\frac{1}{3}}, F_{\frac{2}{3}}, F_1, F_{\frac{4}{3}}, \ldots$$

in which the interval is $\frac{1}{3}\omega$. This process is called *interpolation to thirds*. To interpolate *to fifths*, we let $m = \frac{1}{5}$, thus introducing *four* functions between every two adjacent terms of the original series. We then have the tabular values of

$$F_0, F_{\frac{1}{5}}, F_{\frac{2}{5}}, F_{\frac{3}{5}}, F_{\frac{4}{5}}, F_1, F_{\frac{6}{5}}, \ldots$$

the interval being $\frac{1}{5}\omega$.

More generally, let us take $m = \frac{1}{k}$, where k is a positive odd integer; we thus introduce $k - 1$ equidistant values of the function between every two adjacent terms of the given series. The resulting series will therefore be

$$F_0, F_m, F_{2m}, F_{3m}, \ldots F_{(k-1)m}, F_1, F_{1+m}, \ldots$$

in which the argument interval is $m\omega$, or $\frac{\omega}{k}$. Now, the two adjacent functions of this interpolated series, which, as a pair, fall *midway* between F_0 and F_1, are

$$F_{\left(\frac{k-1}{2}\right)m} \quad \text{and} \quad F_{\left(\frac{k+1}{2}\right)m}$$

that is

$$F_{\left(\frac{1-m}{2}\right)} \quad \text{and} \quad F_{\left(\frac{1+m}{2}\right)}$$

Hence, if we put

$$\delta_{\frac{1}{2}}' = F_{\left(\frac{1+m}{2}\right)} - F_{\left(\frac{1-m}{2}\right)} \tag{131}$$

it follows that $\delta_{\frac{1}{2}}'$ is the value of the *first difference* of the *interpolated* series which falls on the line *midway* between F_0 and F_1; we shall designate this quantity a *middle first difference* of the required series. If we now let

$$\frac{1+m}{2} = n \tag{132}$$

we have

$$\frac{1-m}{2} = 1-n$$

and (131) becomes

$$\delta_{\frac{1}{2}}' = F_n - F_{1-n} \tag{133}$$

Hence, to express $\delta_{\frac{1}{2}}'$ in terms of the differences of the *given* series, we have only to express the values of F_n and F_{1-n} by BESSEL'S Formula; thus, abbreviating coefficients, we have, as in (113),

$$F_n = F_0 + na_1 + Bb + Cc_1 + Dd + Ee_1 + \ldots \tag{134}$$

Also, by virtue of the property of these coefficients established in §41, we have

$$F_{1-n} = F_0 + (1-n)a_1 + Bb - Cc_1 + Dd - Ee_1 + \ldots \tag{135}$$

The difference of these equations gives

$$\delta_{\frac{1}{2}}' = F_n - F_{1-n} = (2n-1)a_1 + 2Cc_1 + 2Ee_1 + \ldots \tag{136}$$

Now, by (132), we have

$$n = \frac{1+m}{2}$$

hence, from (111), we find

$$C = \tfrac{1}{6}n(n-1)(n-\tfrac{1}{2}) = \frac{m}{48}(m^2-1)$$

$$E = \tfrac{1}{120}(n+1)n(n-1)(n-2)(n-\tfrac{1}{2}) = \tfrac{1}{20}(n+1)(n-2)C = \frac{m}{3840}(m^2-1)(m^2-9)$$

Substituting these values of n, C, and E in (136), we obtain the formula

$$\delta_{\frac{1}{2}}' = ma_1 + \frac{m}{24}(m^2-1)\,c_1 + \frac{m}{1920}(m^2-1)(m^2-9)\,e_1 + \ldots \quad (137)$$

by which the *middle first differences* may be computed in any case, provided $\frac{1}{m}$ is a positive odd integer.

Let us now consider the schedule below :

T	$F(T)$	δ'	δ''	δ'''	Δ'	Δ''	Δ'''	Δ^{iv}	Δ^{v}
$t-\omega$	F_{-1}		δ''_{-1}			b'		d'	
$t-\omega+m\omega$	. .	. .	. .	. .					
. . . .	. .	$\delta'_{-\frac{1}{2}}$	. .	$\delta'''_{-\frac{1}{2}}$	a'		c'		e'
. . . .	. .	. .	. .	. .					
$t-m\omega$	. .	. .	. .	. .					
t	F_0	. .	δ_0''	. .		b_0		d_0	
$t+m\omega$	. .	. .	. .	. .					
. . . .	. .	$\delta_{\frac{1}{2}}'$	. .	$\delta_{\frac{1}{2}}'''$	a_1		c_1		e_1
. . . .	. .	. .	. .	. .					
$t+\omega-m\omega$	. .	. .	. .	. .					
$t+\omega$	F_1	. .	δ_1''	. .		b_1		d_1	

The quantities are here arranged in a manner somewhat similar to the schedule of §50. The given functions, F_{-1}, F_0, F_1, ..., are separated, successively, by $k-1$ blank lines or spaces, for the subsequent entry of the interpolated values. The columns δ', δ'', and δ''' are also reserved for the differences of the interpolated series ; and the differences of the given functions are written to the right, in columns Δ' to Δ^{v}.

The value of $\delta_{\frac{1}{2}}'$ is now computed by (137) from the differences a_1, c_1, and e_1, which stand opposite. In like manner, $\delta'_{-\frac{1}{2}}$ is computed from the differences a', c', and e'; $\delta_{\frac{3}{2}}'$, from a_2, c_2, and e_2; and so on. We thus obtain a series of *middle first differences*, which are tabulated under δ' in the schedule above.

Now it is clear that if we should interpolate the $k-1$ intermediate terms between $\delta'_{-\frac{1}{2}}$ and $\delta_{\frac{1}{2}}'$, between $\delta_{\frac{1}{2}}'$ and $\delta_{\frac{3}{2}}'$, etc., the resulting series would constitute the consecutive first differences of the *interpolated* series $F(T)$; the required functions would then be formed by successive additions of these differences. The problem of interpolating the given series $F(T)$ is thus virtually reduced to that of interpolating the *computed* values of δ' in *precisely the same manner.*

Now, let δ_0'' denote the second difference of the *interpolated* series F, which stands opposite F_0; δ_1'', the second difference opposite F_1; etc. It follows that δ_0'' is the *middle first difference* of the *interpolated* series δ', which falls between $\delta'_{-\frac{1}{2}}$ and $\delta'_{\frac{1}{2}}$; δ_1'', that falling between $\delta_{\frac{1}{2}}'$ and $\delta_{\frac{3}{2}}'$; and so on. Hence, we may find $\delta_0'', \delta_1'', \delta_2'', \ldots$ from the computed series $\delta'_{-\frac{1}{2}}, \delta_{\frac{1}{2}}', \delta_{\frac{3}{2}}', \ldots$, in precisely the manner that the latter quantities are derived from $F_{-1}, F_0, F_1, \ldots$; that is, by application of the general formula (137), *mutatis mutandis.* For this purpose, we must form the differences of the computed series

$$\delta'_{-\frac{1}{2}},\ \delta_{\frac{1}{2}}',\ \delta_{\frac{3}{2}}',\ \ldots$$

Accordingly, let us put, for brevity,

$$M = \frac{m}{24}(m^2-1) \quad , \quad M' = \frac{m}{1920}(m^2-1)(m^2-9) \tag{138}$$

and (137) becomes

$$\delta_{\frac{1}{2}}' = ma_1 + Mc_1 + M'e_1 \tag{139}$$

provided differences beyond Δ^{v} are disregarded. We now form a table of the quantities $\delta'_{-\frac{1}{2}}, \delta_{\frac{1}{2}}', \delta_{\frac{3}{2}}', \ldots$, and their differences, as follows:

Function, $=\delta'$	1st Diff.	2d Diff.	3d	4th
$\delta'_{-\frac{1}{2}} = ma' + Mc' + M'e'$	$mb' + Md'$	$mc' + Me'$	md'	me'
$\delta_{\frac{1}{2}}' = ma_1 + Mc_1 + M'e_1$	$mb_0 + Md_0$	$mc_1 + Me_1$	md_0	me_1
$\delta_{\frac{3}{2}}' = ma_2 + Mc_2 + M'e_2$	$mb_1 + Md_1$	$mc_2 + Me_2$	md_1	me_2

Whence, applying the general formula (139) to the quantities of this table, we obtain

$$\delta_0'' = m(mb_0 + Md_0) + M(md_0) = m^2b_0 + 2Mmd_0$$

or, by (138),

$$\delta_0'' = m^2b_0 + \frac{m^2}{12}(m^2-1)d_0 \tag{140}$$

by which the quantities $\delta''_{-1}, \delta_0'', \delta_1'', \ldots$ of the former schedule are computed from the differences Δ'' and Δ^{iv} which stand opposite.

Again, we may suppose that the intermediate values of δ'' have been interpolated between the computed values $\delta''_{-1}, \delta_0'', \delta_1'', \ldots$; this completed series δ'' constitutes the consecutive second differences of the *interpolated* series $F(T)$. Finally, we shall denote by $\delta'''_{\frac{1}{2}}$ the third difference of the interpolated series F, which stands opposite $\delta'_{\frac{1}{2}}$ in the given schedule. The quantity $\delta'''_{\frac{1}{2}}$ is therefore the middle first difference of the completed series δ'', which falls between δ''_0 and δ''_1; it bears the same relation to δ''_0 and δ''_1, that $\delta'_{\frac{1}{2}}$ bears to F_0 and F_1. Hence, to find $\delta'''_{\frac{1}{2}}$, let us put

$$M'' = \frac{m^2}{12}(m^2-1)$$

and (140) becomes

$$\delta''_0 = m^2 b_0 + M'' d_0 \qquad (141)$$

The differences of $\delta''_{-1}, \delta''_0, \delta''_1, \ldots.$ are therefore as follows:

Function, $= \delta''$	1st Diff.	2d	3d
$\delta''_{-1} = m^2 b' + M'' d'$		$m^2 d'$	
	$m^2 c' + M'' e'$		$m^2 e'$
$\delta''_0 = m^2 b_0 + M'' d_0$		$m^2 d_0$	
	$m^2 c_1 + M'' e_1$		$m^2 e_1$
$\delta''_1 = m^2 b_1 + M'' d_1$		$m^2 d_1$	

Whence, applying (as above) the general formula (139), we find

$$\delta'''_{\frac{1}{2}} = m(m^2 c_1 + M'' e_1) + M(m^2 e_1) = m^3 c_1 + (mM'' + m^2 M)\, e_1$$

Substituting the values of M and M'', we have

$$\delta'''_{\frac{1}{2}} = m^3 c_1 + \frac{m^3}{8}(m^2-1)\, e_1 \qquad (142)$$

In practice, the values of δ^{iv} and δ^{v} are never required, and in many cases the column δ''' is not necessary. Supposing, however, that we have computed the (nearly constant) values of $\delta'''_{-\frac{1}{2}}, \delta'''_{\frac{1}{2}}, \delta'''_{\frac{3}{2}}, \ldots$ by (142), the intermediate terms are then written in by mere *inspection.* We thus complete the column δ''',—the consecutive third differences of the required series $F(T)$. Having also computed the quantities $\delta''_0, \delta''_1, \delta''_2, \ldots.$ and $\delta'_{-\frac{1}{2}}, \delta'_{\frac{1}{2}}, \delta'_{\frac{3}{2}}, \ldots.$, we complete the columns δ'' and δ', and hence, also, the interpolated series $F(T)$, by successive additions.

We now bring together the formulae for $\delta'_{\frac{1}{2}}, \delta''_0$, and $\delta'''_{\frac{1}{2}}$, in the order computed in practice, as follows:

$$\left.\begin{aligned}
\delta'''_{\frac{1}{2}} &= m^3 c_1 + \frac{m^3}{8}(m^2-1)\, c_1 \\
\delta''_0 &= m^2 b_0 + \frac{m^2}{12}(m^2-1) d_0 \\
\delta'_{\frac{1}{2}} &= m a_1 + \frac{m}{24}(m^2-1)\, c_1 + \frac{m}{1920}(m^2-1)(m^2-9)\, e_1
\end{aligned}\right\} \qquad (143)$$

which serve to reduce the tabular interval to m times its original value, m being the reciprocal of a positive odd integer. It will be observed that the differences required in computing each of the quantities δ are always found on the same line with that quantity.

52. *Interpolation to Thirds.*—For this purpose, we take $m = \frac{1}{3}$ in the formulae (143), and find

$$\left.\begin{aligned} \delta_{\frac{1}{2}}''' &= \tfrac{1}{27}\, c_1 - \tfrac{1}{243}\, e_1 \\ \delta_0'' &= \tfrac{1}{9}\, b_0 - \tfrac{2}{243}\, d_0 \\ \delta_{\frac{1}{2}}' &= \tfrac{1}{3}\, a_1 - \tfrac{1}{81}\, c_1 + \tfrac{1}{729}\, e_1 \end{aligned}\right\} \qquad (144)$$

These formulae are more conveniently computed in the form

$$\left.\begin{aligned} \delta_{\frac{1}{2}}''' &= \tfrac{1}{27}\,(c_1 - \tfrac{1}{9}\, e_1) \\ \delta_0'' &= \tfrac{1}{9}\,(b_0 - \tfrac{2}{27}\, d_0) \\ \delta_{\frac{1}{2}}' &= \tfrac{1}{3}\,(a_1 - \delta_{\frac{1}{2}}''') \end{aligned}\right\} \qquad (145)$$

EXAMPLE.—Given the value of log tan T for every third degree of T from 27° to 48°, inclusive: find the function for every degree between 33° and 42°.

According to the precepts of the last section, we arrange the work as follows:

T	Log tan T	δ'	δ''	δ'''	Δ'	Δ''	Δ'''	Δ^{iv}
27°	**9.70717**							
					+5427			
30	**9.76144**					−319		
				+3.1	5108		+85	
				3.0				
33	**9.81252**		**−25.9**			234		−14
		+1646.9		2.8				
34	9.82899		23.1					
		1623.8		**2.6**	4874		71	
35	9.84523		20.5					
		1603.3		2.5				
36	**9.86126**		**18.0**			163		12
		1585.3		2.3				
37	9.87711		15.7					
		1569.6		**2.2**	4711		59	
38	9.89281		13.5					
		1556.1		2.0				
39	**9.90837**		**11.5**			104		6
		1544.6		2.0				
40	9.92382		9.5					
		1535.1		**1.9**	4607		53	
41	9.93917		7.6					
		+1527.5		1.9				
42	**9.95444**		**− 5.7**			− 51		− 2
				1.9				
				+1.9	4556		+51	
45	**0.00000**					0		
					+4556			
48	**0.04556**							

The heavy type shows at a glance the given functions, and likewise the computed middle differences. We observe that it is here necessary to compute five values of δ''', four values of δ'', and only three of δ'. These quantities are computed to one more than the number of decimals given in $F(T)$, to avoid accumulation of any appreciable error in the final additions. Having obtained for δ''' the series

$$+3.1 \qquad 2.6 \qquad 2.2 \qquad 1.9 \qquad +1.9$$

the intermediate terms are readily inserted, as shown above; it is necessary, however, to see that the completed series δ''' is consistent with the *computed* values of δ''. Thus we must have

$$\begin{aligned} 2.8 + 2.6 + 2.5 &= -(18.0-25.9) = +7.9 \\ 2.3 + 2.2 + 2.0 &= -(11.5-18.0) = +6.5 \\ 2.0 + 1.9 + 1.9 &= -(\ 5.7-11.5) = +5.8 \end{aligned}$$

If these relations are not satisfied exactly on first trial, the interpolated values of δ''' must be adjusted to fulfill the necessary conditions.

The column δ'' is now completed by successive additions of the quantities δ'''. Again, it is necessary to see that the completed series δ'' agrees with the computed values of δ'. For we must have

$$-(20.5+18.0+15.7) = 1569.6 - 1623.8 = -54.2, \text{ etc.}$$

Since these relations are seldom exact in the beginning, the provisional values of δ'' will usually require slight alterations.

From the final series δ'', we obtain δ' by successive additions. As before, an agreement must subsist between the values of δ' and the given set of functions; that is, between δ' and Δ'. Thus we should have

$$\Sigma\delta' = 1646.9 + 1623.8 + 1603.3 = +4874.0 = \Delta', \text{ etc.}$$

In the latter case, however, a discrepancy not exceeding four or five units in the added decimal may be tolerated. Our final series δ' is therefore satisfactory; whence we obtain by successive additions the required values of log tan T.

53. *Interpolation to Fifths.* — Taking $m = \frac{1}{5}$ in the formulae (143), we obtain

$$\left.\begin{aligned} \delta_{\frac{1}{2}}''' &= \tfrac{1}{125}(c_1 - \tfrac{3}{25}e_1) \\ \delta_0'' &= \tfrac{1}{25}(b_0 - \tfrac{2}{25}d_0) \\ \delta_{\frac{1}{2}}' &= \tfrac{1}{5}\{a_1 - \tfrac{1}{25}(c_1 - \tfrac{14}{125}e_1)\} \end{aligned}\right\} \qquad (146)$$

In practice it will suffice to put $\frac{1}{10}c_1$ for both $\frac{3}{25}e_1$ and $\frac{14}{125}e_1$; the formulae (146) then become, very approximately,

$$\left.\begin{aligned}\delta_{\frac{1}{2}}''' &= \tfrac{1}{125}\left(c_1 - \tfrac{1}{5}\,e_1\right)\\ \delta_0'' &= \tfrac{1}{25}\left(b_0 - \tfrac{2}{25}\,d_0\right)\\ \delta_{\frac{1}{2}}' &= \tfrac{1}{5}\,a_1 - \delta_{\frac{1}{2}}'''\end{aligned}\right\}\qquad (147)$$

EXAMPLE. — The following ephemeris gives the moon's R.A. for every ten hours. Obtain the value for every second hour, from Sept. 23ᵈ 20ʰ to Sept. 25ᵈ 12ʰ, inclusive.

The details of the computation are as follows :

Date, 1898	Moon's R.A.	δ'	δ''	δ'''	Δ'	Δ''	Δ'''	Δ^{iv}
d h	h m s	m s	s	s	m s	s	s	s
Sept. 23 0	18 24 26.4							
					+25 31.1			
Sept. 23 10	18 49 57.5					−20.1		
				−.034	25 11.0		−4.2	
				32				
				30				
Sept. 23 20	19 15 8.5		−0.976			24.3		+1.2
		+4 59.39		28				
23 22	19 20 7.9		1.004					
		58.39		26				
24 0	19 25 6.3		1.030					
		4 57.36		.024	24 46.7		3.0	
24 2	19 30 3.7		1.054					
		56.31		23				
24 4	19 35 0.0		1.077					
		55.23		20				
Sept. 24 6	19 39 55.2		1.097			27.3		1.5
		54.13		17				
24 8	19 44 49.3		1.114					
		53.02		14				
24 10	19 49 42.3		1.128					
		4 51.89		.012	24 19.4		1.5	
24 12	19 54 34.2		1.140					
		50.75		09				
24 14	19 59 25.0		1.149					
		49.60		07				
Sept. 24 16	20 4 14.6		1.156			28.8		1.4
		48.44		05				
24 18	20 9 3.0		1.161					
		47.28		03				
24 20	20 13 50.3		1.164					
		4 46.12		−.001	23 50.6		−0.1	
24 22	20 18 36.4		1.165					
		44.95		+.002				
25 0	20 23 21.4		1.163					
		43.79		03				
Sept. 25 2	20 28 5.2		1.160			28.9		1.1
		42.63		06				
25 4	20 32 47.8		1.154					
		41.47		07				
25 6	20 37 29.3		1.147					
		4 40.33		.008	23 21.7		+1.0	
25 8	20 42 9.6		1.139					
		39.19		09				
25 10	20 46 48.8		1.130					
		+4 38.06		11				
Sept. 25 12	20 51 26.9		−1.119			27.9		+0.9
				12				
				14				
				+.015	22 53.8		+1.9	
Sept. 25 22	21 14 20.7					−26.0		
					+22 27.8			
Sept. 26 8	21 36 48.5							

Here we extend the computation of δ''' and δ'' *two places* of decimals; one of which is dropped in computing δ', and the other in forming the required functions. The principle and method being the same as in the last example, further explanation is unnecessary.

54. *Order of Interpolation to Follow, when a Series Requires Successive Interpolation to Halves, Thirds, etc.* — When a table of functions is to be interpolated, successively, one or more times to *halves*, and also to *thirds* and *fifths*, the easiest method is to proceed in the order named. Thus, if the interval of the original series is ω, and that of the final table is ω', we may suppose the relation of these quantities to be —

$$\omega = 2^k . 3^l . 5^m . \omega'$$

where k, l, and m are integers. It will then be found most expedient, first, to interpolate to halves, k times; then to thirds, l times; and finally to fifths, m times.

For example, F being given for every *degree*, and required for every *minute* of arc, we should first interpolate to 30′, then to 15′, then to 5′, and finally to every minute of arc.

55. *To Interpolate with a Constant Interval n, an Entire Series of Functions.* — Let the given series, with its differences, be as follows:

T	$F(T)$	Δ'	Δ''	Δ'''	Δ^{iv}
t	F_0		b_0		d_0
		a_1		c_1	
$t+\omega$	F_1		b_1		d_1
		a_2		c_2	
$t+2\omega$	F_2		b_2		d_2
		a_3		c_3	
$t+3\omega$	F_3		b_3		d_3
		a_4		c_4	
$t+4\omega$	F_4		b_4		d_4

It is required to interpolate the values of F_n, F_{1+n}, F_{2+n}, F_{3+n}, ... These functions evidently form a new series having the same interval as the old. Let us denote this new series by $[F]$; also, let the differences of $[F]$, denoted by $[\Delta']$, $[\Delta'']$, $[\Delta''']$, ..., be taken as shown in the table below:

T	$[F]$	$[\Delta']$	$[\Delta'']$	$[\Delta''']$	$[\Delta^{iv}]$
$t+n\omega$	F_n		β_0		δ_0
$t+(1+n)\,\omega$	F_{1+n}	α_1	β_1	γ_1	δ_1
$t+(2+n)\,\omega$	F_{2+n}	α_2	β_2	γ_2	δ_2
$t+(3+n)\,\omega$	F_{3+n}	α_3	β_3	γ_3	δ_3
$t+(4+n)\,\omega$	F_{4+n}	α_4	β_4	γ_4	δ_4

Now, it was shown in §22 that differences of *any* order may be expressed in terms of the tabular functions. Thus, in particular, we obtain from the given series F,

$$\left.\begin{aligned} c_1 &= F_2 - 3F_1 + 3F_0 - F_{-1} \equiv \Psi(t) \\ c_2 &= F_3 - 3F_2 + 3F_1 - F_0 = \Psi(t+\omega) \\ c_3 &= F_4 - 3F_3 + 3F_2 - F_1 = \Psi(t+2\omega) \\ &\ldots\ldots\ldots\ldots \end{aligned}\right\} \qquad (148)$$

where $\Psi(t)$ denotes, for brevity, the function of t expressed by

$F_2 - 3F_1 + 3F_0 - F_{-1}$; that is, $\Psi(t) \equiv F(t+2\omega) - 3F(t+\omega) + 3F(t) - F(t-\omega)$

Again, in like manner, the *interpolated* series $[F]$ gives

$$\left.\begin{aligned} \gamma_1 &= F_{2+n} - 3F_{1+n} + 3F_n - F_{-1+n} = \Psi(t+n\omega) \\ \gamma_2 &= F_{3+n} - 3F_{2+n} + 3F_{1+n} - F_n = \Psi(t+\omega+n\omega) \\ &\ldots\ldots\ldots\ldots \end{aligned}\right\} \qquad (149)$$

It follows, then, that the series $[\Delta''']$ is simply the series Δ''' interpolated forward with the constant interval n. Moreover, since the above reasoning is perfectly general, this relation holds for *any order* of differences.

Hence, to perform the required interpolation of the series $F(T)$, that is, to obtain the series $[F]$, we have only to interpolate forward each value of Δ'' with the constant interval n, thus forming the column $[\Delta'']$. This process is obviously brief and simple. Then, if we compute *occasional* values of $[\Delta']$, and also of $[F]$, we readily complete the required table by successive additions, as in the preceding problems.

Example.—To illustrate the process, we tabulate the "Latitude Reduction" for every fourth degree of latitude (σ) from 30° to 82°, and thence derive the series for $\varphi = 35°, 39°, 43°, \ldots\ldots 75°$. The work is arranged as follows:

φ	$\varphi-\varphi'$	Δ'	Δ''	Δ'''	Δ^{iv}	φ	$[\varphi-\varphi']$	$[\Delta']$	$[\Delta'']$
°	″					°	″		
30	605.56								
		+43.04							
34	648.60		−12.58			**35**	**657.422**		
		30.46		−0.60				**+27.212**	
38	679.06		13.18		+.24	39	684.634		**−13.294**
		17.28		0.36				13.918	
42	696.34		13.54		.27	43	698.552		**13.587**
		+ 3.74		−0.09				+ 0.331	
46	700.08		13.63		.26	47	698.883		**13.612**
		− 9.89		+0.17				−13.281	
50	690.19		13.46		.27	51	685.602		**13.376**
		23.35		0.44				26.657	
54	666.84		13.02		.26	**55**	**658.945**		**12.868**
		36.37		0.70				**39.525**	
58	630.47		12.32		.23	59	619.420		**12.110**
		48.69		0.93				51.635	
62	581.78		11.39		.24	63	567.785		**11.118**
		60.08		1.17				62.753	
66	521.70		10.22		.19	67	505.032		**9.897**
		70.30		1.36				72.650	
70	451.40		8.86		.19	71	432.382		**− 8.488**
		79.16		1.55				**−81.138**	
74	372.24		7.31		+.15	**75**	**351.244**		
		86.47		+1.70					
78	285.77		− 5.61						
		−92.08							
82	193.69								

Taking $n = 0.25$, we compute by BESSEL's Formula the values of $\varphi-\varphi'$ for $\varphi = 35°, 55°$, and $75°$, extending the decimal one unit. Similarly, we compute *three* values of $[\Delta']$, and *all* of $[\Delta'']$; the computed quantities being clearly shown by heavier type. Adjusting slightly the series $[\Delta'']$ to conform to the *computed* values of $[\Delta']$, we complete the latter column by successive additions. The values of $[\Delta']$ being found to accord with the computed functions, we complete the entire series as required.

Since the computed intermediate values of $[\Delta']$ and $[F]$ serve only as checks, it is obvious that their positions, as also the intervals of their distribution, are entirely arbitrary. These are details to be decided by the computer's judgement in any given case.

It may occasionally be practicable to extend the process to the computation of $[\Delta''']$.

EXAMPLES.

1. Tabulate the five-place log cosines of 15°, 18°, 21°, 24°, 27°, 30°; from these values interpolate log cos T for $T = 17° 43'$, $23° 8'$, and $28° 15'$, respectively.

2. Given the following table:

T	$F(T)$	T	$F(T)$
	″		″
10	17.31	40	14.16
20	14.68	50	16.34
30	13.62	60	20.18

Compute the values of F for $T = 24.6$, 28.8, 32.3, and 48.5, using either BESSEL's or STIRLING's Formula.

3. Interpolate the required functions of Example 2 by means of a *corrected first difference*, as explained in §§ 44 and 45.

4. What is the maximum error of interpolation in the table of Example 2, supposing that second differences are neglected?

5. Find the correct values of the erroneous functions in the several tables of Example 6, Chap. I, by direct interpolation, as explained in §§ 46 and 47.

6. Given the following twelve-hour ephemeris of lunar distances of *Spica*:

Date 1898	L. D. of *Spica*	Date 1898	L. D. of *Spica*
d	° ′ ″	d	° ′ ″
July 1.0	43 24 9	July 3.0	73 35 46
1.5	50 52 0	3.5	81 12 52
2.0	58 24 0	4.0	88 48 56
July 2.5	65 59 1	July 4.5	96 22 40

Interpolate the series *twice* to halves; the first result to include the values from July $1^d.5$ to $4^d.0$, and the final three-hour ephemeris to extend from July 2^d 0^h to July 3^d 12^h, inclusive.

7. The ephemeris below gives the sun's true longitude for every third day:

1898	Sun's Longitude	1898	Sun's Longitude
	° ′ ″		° ′ ″
Oct. 7	194 14 35.2	Oct. 16	203 9 32.9
10	197 12 34.2	19	206 8 29.4
Oct. 13	200 10 54.0	Oct. 22	209 7 42.0

Derive from these values a *daily* ephemeris extending from Oct. 10 to Oct. 19, inclusive.

8. The following table contains the heliocentric longitude of *Jupiter* for every 80th day of 1898–99, beginning with Jan. 0, 1898:

Date 1898	Helioc. Long. of *Jupiter*	Date 1898	Helioc. Long. of *Jupiter*
d	° ′ ″	d	° ′ ″
0	178 59 17.9	320	203 10 20.5
80	185 2 24.1	400	209 13 53.8
160	191 5 0.3	480	215 18 35.1
240	197 7 30.9	560	221 24 48.1

Interpolate this table to halves, extending the series from 120^d to 440^d inclusive; designate this forty-day ephemeris, Table *A*. Then interpolate *A* to fifths, denoting the eight-day series by *B*. Let the limits of *B* be 200^d and 320^d, respectively. Retain copies of *A* and *B*.

9. Interpolate (forward) the longitudes of Table *A*, Example 8, with the constant interval $n = 0.20$, by the method of §55. This will furnish an ephemeris for the dates 168^d, 208^d, 248^d, ... 368^d. Compare the longitudes thus found for 208^d, 248^d, and 288^d, with their values in Table *B*, Example 8.

10. Deduce from the general formulae (143), the special formulae for interpolation to *sevenths*. Make an application to the five-figure

logarithms of 47, 54, 61, ... 96, by computing the logarithms of the consecutive numbers between 61 and 75.

11. Show that if the formulae (143) were extended to include the middle differences of order i, we should have (using the symbolic form of notation employed in the analogous formulae (64))

$$\begin{aligned}\delta^i &= (\delta')^i = \left(m\Delta + \frac{m}{24}(m^2-1)\Delta^3 + \frac{m}{1920}(m^2-1)(m^2-9)\Delta^5 + \ldots\ldots\right)^i \\ &= m^i\Delta^i + \frac{im^i}{24}(m^2-1)\Delta^{i+2} + \frac{im^i}{5760}(m^2-1)\{(5i-2)m^2-(5i+22)\}\Delta^{i+4} + \ldots\ldots\end{aligned}$$

in which i may be either odd or even; and where Δ^i, Δ^{i+2}, Δ^{i+4}, ... symbolize the tabular differences which fall upon the same horizontal line with δ^i.

CHAPTER III.

DERIVATIVES OF TABULAR FUNCTIONS.

56. It is often required to find certain numerical values of the differential coefficients of functions either analytically unknown, or complicated in expression. In the majority of such cases the function has been previously tabulated for particular (equidistant) values of the argument. The required derivatives are then readily computed from the *differences* of the tabular functions.

We have already seen that — with certain limitations — particular values of a function, with their differences, practically determine the character and law of that function, thus enabling us to determine intermediate values by interpolation. The trend or law of variation of the function being thus defined by its differences, it is but natural to suppose that the successive *derivatives* are quantities closely related to these *differences*; since the derivatives are themselves direct indices of the character of variation of the function.

57. *Practical Applications.*—The most useful application is in finding the change or *variation* in $F(T)$ corresponding to an increase of *one unit* in T, supposing the *rate* of change in F to remain constant from T to $T+1$, and equal to the actual rate at the instant T; for this quantity is simply the first differential coefficient of $F(T)$ with respect to T, which we shall denote by $F'(T)$.

For example, having observed that a freely falling body describes sixteen feet during the first second of its descent, forty-eight feet the second second, and eighty feet the third, its *velocity* at the end of two seconds is easily found to be sixty-four feet per second. This velocity of sixty-four feet is nothing more than the first differential coefficient of the *space* with respect to the *time*, computed for the instant $2^s.0$: it is the space which would be described during the third second, supposing the action of gravity to have ceased at the end of the second second.

The most frequent and important applications occur in Astronomy. An astronomical ephemeris contains a great variety of tables giving the positions and motions of various heavenly bodies, and of certain points of reference. From the given positions, tabulated for every hour or from day to day, are derived the motions per minute, per hour, or per day, according to circumstances. For instance, the *Nautical Almanac* gives the sun's declination for every Greenwich noon. The *hourly motion* in declination (also given for every noon) is computed from the *differences* of the tabular declinations: its value is the differential coefficient of the tabular function at the date in question.

In the following sections the various formulae employed in computing the derivatives of tabular functions will be derived.

58. *Development of the Required Formulae in General Terms.*— The variables T and n are connected by the fundamental relation

$$T = t + n\omega \tag{150}$$

in which t and ω are constants for a given series. Accordingly, we have hitherto written, under varying circumstances,

$$F(T) \quad , \quad F(t+n\omega) \quad , \quad F_n$$

as equivalent expressions of the same quantity. In like manner, we shall hereafter denote the successive *derivatives* of $F(T)$ by the following equivalent forms:

$$\left.\begin{array}{l} \frac{d}{dT}\left\{F(T)\right\} \equiv F'(T) \equiv F'(t+n\omega) \equiv F'_n \\ \frac{d^2}{dT^2}\left\{F(T)\right\} \equiv F''(T) \equiv F''(t+n\omega) \equiv F''_n \\ \frac{d^3}{dT^3}\left\{F(T)\right\} \equiv F'''(T) \equiv F'''(t+n\omega) \equiv F'''_n \\ \frac{d^4}{dT^4}\left\{F(T)\right\} \equiv F^{\text{iv}}(T) \equiv F^{\text{iv}}(t+n\omega) \equiv F^{\text{iv}}_n \\ \cdots\cdots\cdots\cdots\cdots\cdots \end{array}\right\} \tag{151}$$

When it is convenient to proceed *backwards* from the argument t with the interval n, we shall use the expressions

$$F'_{-n} \equiv F'(t-n\omega) \quad , \quad F''_{-n} \equiv F''(t-n\omega) \quad , \quad F'''_{-n} \equiv F'''(t-n\omega) \quad , \quad \ldots\ldots \tag{152}$$

Now, by means of any one of the fundamental formulae of interpolation, we may express F_n in the form

$$F_n = F_0 + na + Bb + Cc + Dd + Ee + \ldots \quad (153)$$

where, in any given case, $a, b, c, \ldots$ are known differences; and where $B, C, D, \ldots$ are definite functions of n. Let the successive derivatives of $B, C, D, \ldots$, taken with respect to n, be denoted by

$$\begin{array}{l} B', B'', B''', \ldots \\ C', C'', C''', \ldots \\ D', D'', D''', \ldots \\ E', E'', E''', \ldots \\ \ldots\ldots\ldots \end{array}$$

Then, observing that the coefficient of $\Delta^{(i)}$ is always of the degree i in n, we have

$$\left.\begin{array}{llll} \frac{dB}{dn} = B' & \frac{dC}{dn} = C' & \frac{dD}{dn} = D' & \frac{dE}{dn} = E' \\ \frac{d^2B}{dn^2} = B'' & \frac{d^2C}{dn^2} = C'' & \frac{d^2D}{dn^2} = D'' & \frac{d^2E}{dn^2} = E'' \\ \frac{d^3B}{dn^3} = 0 & \frac{d^3C}{dn^3} = C''' & \frac{d^3D}{dn^3} = D''' & \frac{d^3E}{dn^3} = E''' \\ & \frac{d^4C}{dn^4} = 0 & \frac{d^4D}{dn^4} = D^{\text{iv}} & \frac{d^4E}{dn^4} = E^{\text{iv}} \\ & & \frac{d^5D}{dn^5} = 0 & \frac{d^5E}{dn^5} = E^{\text{v}} \\ & & & \frac{d^6E}{dn^6} = 0 \end{array}\right\} \quad (154)$$

Reverting to (151), we have

$$F'_n = \frac{dF_n}{dT} = \frac{dF_n}{dn} \cdot \frac{dn}{dT} \quad (155)$$

From (150) we derive

$$\frac{dn}{dT} = \frac{1}{\omega} \quad (156)$$

whence

$$F'_n = \frac{1}{\omega} \cdot \frac{dF_n}{dn} \quad (157)$$

In like manner we obtain

$$\left.\begin{aligned}
F''_n &= \frac{dF'_n}{dT} = \frac{dF'_n}{dn}\cdot\frac{dn}{dT} = \frac{1}{\omega^2}\cdot\frac{d^2F_n}{dn^2}\\
F'''_n &= \frac{dF''_n}{dT} = \frac{dF''_n}{dn}\cdot\frac{dn}{dT} = \frac{1}{\omega^3}\cdot\frac{d^3F_n}{dn^3}\\
F^{\text{iv}}_n &= \frac{dF'''_n}{dT} = \frac{dF'''_n}{dn}\cdot\frac{dn}{dT} = \frac{1}{\omega^4}\cdot\frac{d^4F_n}{dn^4}\\
F^{\text{v}}_n &= \frac{dF^{\text{iv}}_n}{dT} = \frac{dF^{\text{iv}}_n}{dn}\cdot\frac{dn}{dT} = \frac{1}{\omega^5}\cdot\frac{d^5F_n}{dn^5}\\
&\ldots\ldots\ldots\ldots\ldots
\end{aligned}\right\} \tag{158}$$

Therefore, using (153) and (154), we find

$$\left.\begin{aligned}
F'_n &= \frac{1}{\omega}\,(a+B'b+C'c+D'd+E'e+\ldots\ldots)\\
F''_n &= \frac{1}{\omega^2}\,(B''b+C''c+D''d+E''e+\ldots\ldots)\\
F'''_n &= \frac{1}{\omega^3}\,(C'''c+D'''d+E'''e+\ldots\ldots)\\
F^{\text{iv}}_n &= \frac{1}{\omega^4}\,(D^{\text{iv}}d+E^{\text{iv}}e+\ldots\ldots)\\
F^{\text{v}}_n &= \frac{1}{\omega^5}\,(E^{\text{v}}e+\ldots\ldots)\\
&\ldots\ldots\ldots
\end{aligned}\right\} \tag{159}$$

which are the general formulae for computing the derivatives of $F(T)$ in terms of the tabular differences.

To derive the formulae for F'_{-n}, F''_{-n}, F'''_{-n}, ... , that is, to find the successive derivatives of $F(t-n\omega)$, we have only to alter slightly certain details of the preceding development, as follows:

(1) For equation (153) must be substituted the corresponding expression for F_{-n}, which has the form*

$$F_{-n} = F_0 - na + B\beta - C\gamma + D\delta - E\epsilon + \ldots\ldots \tag{160}$$

where α, β, γ, ... are, in general, different from the differences a, b, c, ... of (153).

(2) In the present case, we have

$$T = t - n\omega$$

and therefore

$$\frac{dn}{dT} = -\frac{1}{\omega}$$

which must be substituted for equation (156) above.

* Compare (75), (105) and (111*a*) with (73), (104) and (111), respectively.

Introducing these changes, and operating as before, we obtain the required formulae, namely,

$$\left.\begin{aligned} F'_{-n} &= \frac{1}{\omega}\left(a - B'\beta + C'\gamma - D'\delta + E'\epsilon - \ldots\right) \\ F''_{-n} &= \frac{1}{\omega^2}\left(B''\beta - C''\gamma + D''\delta - E''\epsilon + \ldots\right) \\ F'''_{-n} &= \frac{1}{\omega^3}\left(C'''\gamma - D'''\delta + E'''\epsilon - \ldots\right) \\ F^{\text{iv}}_{-n} &= \frac{1}{\omega^4}\left(D^{\text{iv}}\delta - E^{\text{iv}}\epsilon + \ldots\right) \\ F^{\text{v}}_{-n} &= \frac{1}{\omega^5}\left(E^{\text{v}}\epsilon - \ldots\right) \\ &\ldots\ldots\ldots \end{aligned}\right\} \quad (161)$$

It now remains to apply (159) and (161) specifically to each of the several formulae of interpolation, of which (153) is the general type. It is obvious that a particular set of coefficients, $B', B'', \ldots,$ $C', C'', \ldots,$ etc., will result in each case.

59. *To Compute Derivatives of $F(T)$ at or near the Beginning of a Series.*—The formulae adapted to this purpose are derived from NEWTON's Formula of interpolation (73), which is—

$$F_n = F_0 + na_0 + Bb_0 + Cc_0 + Dd_0 + Ee_0 + \ldots \quad (162)$$

where

$$\left.\begin{aligned} B &= \frac{n(n-1)}{\lfloor 2} = \frac{n^2}{2} - \frac{n}{2} \\ C &= \frac{n(n-1)(n-2)}{\lfloor 3} = \frac{n^3}{6} - \frac{n^2}{2} + \frac{n}{3} \\ D &= \frac{n(n-1)(n-2)(n-3)}{\lfloor 4} = \frac{n^4}{24} - \frac{n^3}{4} + \frac{11}{24}n^2 - \frac{n}{4} \\ E &= \frac{n(n-1)\ldots(n-4)}{\lfloor 5} = \frac{n^5}{120} - \frac{n^4}{12} + \frac{7}{24}n^3 - \frac{5}{12}n^2 + \frac{n}{5} \\ &\ldots\ldots\ldots\ldots\ldots \end{aligned}\right\} \quad (163)$$

Differentiating these expressions successively with respect to n, as indicated in (154), and substituting the resulting values of $B', B'', \ldots,$ $C', C'', \ldots,$ etc., in the general formulae (159), we obtain

$$\left.\begin{aligned}
F'(t+n\omega) &= \frac{1}{\omega}\Big(a_0 + (n-\tfrac{1}{2})\,b_0 + (\tfrac{n^2}{2} - n + \tfrac{1}{3})c_0 + (\tfrac{n^3}{6} - \tfrac{3}{4}n^2 + \tfrac{11}{12}n - \tfrac{1}{4})\,d_0 \\
&\qquad + (\tfrac{n^4}{24} - \tfrac{n^3}{3} + \tfrac{7}{8}n^2 - \tfrac{5}{6}n + \tfrac{1}{5})e_0 + \;.\;.\;.\;.\Big) \\
F''(t+n\omega) &= \frac{1}{\omega^2}\Big(b_0 + (n-1)\,c_0 + (\tfrac{n^2}{2} - \tfrac{3}{2}n + \tfrac{11}{12})d_0 + (\tfrac{n^3}{6} - n^2 + \tfrac{7}{4}n - \tfrac{5}{6})\,e_0 + \;.\;.\Big) \\
F'''(t+n\omega) &= \frac{1}{\omega^3}\Big(c_0 + (n-\tfrac{3}{2})\,d_0 + (\tfrac{n^2}{2} - 2n + \tfrac{7}{4})e_0 + \;.\;.\;.\;.\Big) \\
F^{\text{iv}}(t+n\omega) &= \frac{1}{\omega^4}\Big(d_0 + (n-2)\,e_0 + \;.\;.\;.\;.\Big) \\
F^{\text{v}}(t+n\omega) &= \frac{1}{\omega^5}\Big(e_0 + \;.\;.\;.\;.\Big) \\
&.\;.\;.\;.\;.\;.\;.\;.\;.\;.
\end{aligned}\right\} \quad (164)$$

These formulae determine the derivatives of $F(T)$ for any or all values of T between t and $t+\omega$, according as we assign different values to n. As in preceding applications, n is always a positive proper fraction.

When, as is frequently the case, derivatives are required for some *tabular* value of the argument, say t, we have only to make $n = 0$ in (164); we thus derive the following simple expressions:

$$\left.\begin{aligned}
F'(t) &= \frac{1}{\omega}\,(a_0 - \tfrac{1}{2}\,b_0 + \tfrac{1}{3}\,c_0 - \tfrac{1}{4}\,d_0 + \tfrac{1}{5}\,e_0 - \;.\;.\;.\;.\,) \\
F''(t) &= \frac{1}{\omega^2}\,(b_0 - c_0 + \tfrac{11}{12}\,d_0 - \tfrac{5}{6}\,e_0 + \;.\;.\;.\;.\,) \\
F'''(t) &= \frac{1}{\omega^3}\,(c_0 - \tfrac{3}{2}\,d_0 + \tfrac{7}{4}\,e_0 - \;.\;.\;.\;.\,) \\
F^{\text{iv}}(t) &= \frac{1}{\omega^4}\,(d_0 - 2e_0 + \;.\;.\;.\;.\,) \\
F^{\text{v}}(t) &= \frac{1}{\omega^5}\,(e_0 - \;.\;.\;.\;.\,)
\end{aligned}\right\} \quad (165)$$

The differences employed in (164) and (165) must be taken according to the schedule on page 3, as in direct applications of NEWTON's Formula.

The formulae (165) have already been established in § 18; for it will be observed that (45) and (165) are identical, since in the former $D, D^2, D^3, \ldots$ are used symbolically to denote $\omega F'(t)$, $\omega^2 F''(t)$, $\omega^3 F'''(t), \ldots$

Owing to the special practical importance of the *first* derivative, the coefficients of $F'(t+n\omega)$, namely,

$$\left.\begin{array}{lll} B' = n - \frac{1}{2} & , & D' = \frac{n^3}{6} - \frac{3}{4}n^2 + \frac{11}{12}n - \frac{1}{4} \\ C' = \frac{n^2}{2} - n + \frac{1}{3} & , & E' = \frac{n^4}{24} - \frac{n^3}{3} + \frac{7}{8}n^2 - \frac{5}{6}n + \frac{1}{5} \end{array}\right\} \quad (166)$$

have been tabulated in Table IV for every hundredth of a unit in the argument n. By means of these quantities, we readily compute $F'(t+n\omega)$ from the formula

$$F'(t+n\omega) = \frac{1}{\omega}\left(a_0 + B'b_0 + C'c_0 + D'd_0 + E'e_0\right) \quad (167)$$

The formulae (164), (165), and (167) are especially adapted to the computation of derivatives at or near the beginning of a tabular series. We shall now solve a few examples to illustrate their use.

EXAMPLE I.—From the following table of $F(T) \equiv 0.3T^4 - 2T^2 + 4$, compute $F''(T)$ for $T = 2.8$.

T	$F(T)$	Δ'	Δ''	Δ'''	Δ^{iv}
0	4.0				
		− 3.2			
2	0.8		+ 51.2		
		+ 48.0		+172.8	
4	48.8		224.0		+115.2
		272.0		288.0	
6	320.8		512.0		+115.2
		784.0		+403.2	
8	1104.8		+915.2		
		+1699.2			
10	2804.0				

Here we have

$$t = 2 \qquad \omega = 2 \qquad a_0 = +48.0 \qquad c_0 = +288.0$$
$$T = 2.8 \qquad n = 0.40 \qquad b_0 = +224.0 \qquad d_0 = +115.2$$

Hence, using the second equation of (164), we find

$$\begin{array}{ll} & b_0 = +224.0 \\ C'' = n - 1 = -0.60 & C''c_0 = -172.80 \\ D'' = \frac{n^2}{2} - \frac{3}{2}n + \frac{11}{12} = +0.39\tfrac{2}{3} & D''d_0 = +\ 45.696 \\ \hline & \therefore\ \omega^2 F_n'' = +\ 96.896 \end{array}$$

Whence we obtain

$$F_n'' = 96.896 \div 4 = +24.224$$

This result is easily verified from the known analytical form of the function; thus, since

$$F(T) = 0.3T^4 - 2T^2 + 4$$

we derive

$$F'(T) = 1.2\,T^3 - 4T \quad , \quad F''(T) = 3.6T^2 - 4$$

Substituting $T = 2.8$ in the last equation, we obtain

$$F''(T) = +24.224$$

as found above.

EXAMPLE II. — From the table of the last example, compute $F'(T)$ for $T = 0$.

Here we employ the first of (165). Making $t = 0$, we have

$$a_0 = -3.2 \qquad b_0 = +51.2 \qquad c_0 = +172.8 \qquad d_0 = +115.2$$

We therefore obtain

$$F'(t) = \tfrac{1}{2}\left(-3.2 - \tfrac{51.2}{2} + \tfrac{172.8}{3} - \tfrac{115.2}{4}\right) = 0$$

The result is obviously correct; for we have

$$F'(T) = 1.2T^3 - 4T$$

which vanishes for $T = 0$.

EXAMPLE III. — Given the following table of $F(T) \equiv \sin^2 T$: compute $F''(T)$ for $T = 8°\ 36'$.

T	$F(T) \equiv \sin^2 T$	Δ'	Δ''	Δ'''	Δ^{iv}	Δ^{v}
4°	0.004866					
		+14503				
8	0.019369		+9355			
		23858		−464		
12	0.043227		8891		−174	
		32749		638		+14
16	0.075976		8253		160	
		41002		798		+15
20	0.116978		7455		−145	
		48457		−943		
24	0.165435		+6512			
		+54969				
28	0.220404					

Here we have

$$t = 8° \qquad\qquad T = 8°\ 36'$$

$$\omega = 4° = \frac{\pi}{45} = 0.069813+ \qquad\qquad n = \frac{36}{4 \times 60} = 0.15$$

Taking the coefficients B', C', D' and E' from Table IV with $n = 0.15$, and the differences a_0, b_0, c_0, ... from the given table, we find, in accordance with (167),

		a_0	$= +0.023858$
$B' = -0.35$	$b_0 = +8891$	$B'b_0$	$= -\ 3111.9$
$C' = +0.19458$	$c_0 = -\ 638$	$C'c_0$	$= -\ 124.1$
$D' = -0.12881$	$d_0 = -\ 160$	$D'd_0$	$= +\ 20.6$
$E' = +0.09358$	$e_0 = +\ 15$	$E'e_0$	$= +\ 1.4$

$$\log(\omega F'_n) = 8.314794 \qquad \therefore\ \omega F'_n = +0.020644$$
$$\log \omega = 8.843937$$
$$\log F'_n = 9.470857 \qquad \therefore\ F'_n = +0.295704$$

This result is easily verified by observing that

$$F'(T) = \frac{d}{dT}(\sin^2 T) = \sin 2T$$

which, for $T = 8° 36'$, becomes

$$F'(T) = \sin 17° 12' = 0.295708$$

The former value is thus seen to be very nearly exact.

If the variation in $F(T)$ corresponding to an increase of *one degree* in T were required in the present example, the result would be, simply,

$$F'(T) = 0.020644 \div 4 = +0.005161$$

60. *To Compute Derivatives of $F(T)$ at or near the End of a Series.*—In this case the requisite formulae are derived from NEWTON's Formula for *backward* interpolation (75), namely,

$$F_{-n} = F_0 - na_{-1} + Bb_{-2} - Cc_{-3} + Dd_{-4} - Ee_{-5} + \ldots \qquad (168)$$

where B, C, D, ... have the values given by (163), as before; and where the differences a_{-1}, b_{-2}, c_{-3}, ... are taken according to the schedule below:

T	$F(T)$	Δ'	Δ''	Δ'''	Δ^{iv}	Δ^{v}
$t-5\omega$	F_{-5}		b_{-6}		d_{-7}	
		a_{-5}		c_{-6}		e_{-7}
$t-4\omega$	F_{-4}		b_{-5}		d_{-6}	
		a_{-4}		c_{-5}		e_{-6}
$t-3\omega$	F_{-3}		b_{-4}		d_{-5}	
		a_{-3}		c_{-4}		e_{-5}
$t-2\omega$	F_{-2}		b_{-3}		d_{-4}	
		a_{-2}		c_{-3}		
$t-\omega$	F_{-1}		b_{-2}			
		a_{-1}				
t	F_0					

Comparing (168) with the general formula (160), we have

$$\alpha = a_{-1} \quad , \quad \beta = b_{-2} \quad , \quad \gamma = c_{-3} \quad ,$$

Therefore, substituting the previously determined values of B', B'', ..., C', C'', ..., etc., in the general formulae (161), we obtain

$$\left.\begin{aligned}
F^{\prime}(t-n\omega) &= \frac{1}{\omega}\Big(a_{-1}-(n-\tfrac{1}{2})\,b_{-2}+(\tfrac{n^2}{2}-n+\tfrac{1}{3})\,c_{-3}-(\tfrac{n^3}{6}-\tfrac{3}{4}n^2+\tfrac{11}{12}n-\tfrac{1}{4})\,d_{-4} \\
&\qquad +(\tfrac{n^4}{24}-\tfrac{n^3}{3}+\tfrac{7}{8}n^2-\tfrac{5}{6}n+\tfrac{1}{5})\,e_{-5}-\ldots.\Big) \\
F^{\prime\prime}(t-n\omega) &= \frac{1}{\omega^2}\Big(b_{-2}-(n-1)c_{-3}+(\tfrac{n^2}{2}-\tfrac{3}{2}n+\tfrac{11}{12})d_{-4}-(\tfrac{n^3}{6}-n^2+\tfrac{7}{4}n-\tfrac{5}{6})e_{-5}+\,.\,.\Big) \\
F^{\prime\prime\prime}(t-n\omega) &= \frac{1}{\omega^3}\Big(c_{-3}-(n-\tfrac{3}{2})\,d_{-4}+(\tfrac{n^2}{2}-2n+\tfrac{7}{4})\,e_{-5}-\ldots.\Big) \\
F^{\text{iv}}(t-n\omega) &= \frac{1}{\omega^4}\Big(d_{-4}-(n-2)\,e_{-5}+\ldots.\Big) \\
F^{\text{v}}(t-n\omega) &= \frac{1}{\omega^5}\Big(e_{-5}-\ldots.\Big) \\
&\ldots\ldots\ldots
\end{aligned}\right\} \quad (169)$$

Making $n = 0$ in (169), we have

$$\left.\begin{aligned}
F^{\prime}(t) &= \frac{1}{\omega}\,(a_{-1}+\tfrac{1}{2}\,b_{-2}+\tfrac{1}{3}\,c_{-3}+\tfrac{1}{4}\,d_{-4}+\tfrac{1}{5}\,e_{-5}+\ldots.) \\
F^{\prime\prime}(t) &= \frac{1}{\omega^2}\,(b_{-2}+c_{-3}+\tfrac{11}{12}\,d_{-4}+\tfrac{5}{6}\,e_{-5}+\ldots.) \\
F^{\prime\prime\prime}(t) &= \frac{1}{\omega^3}\,(c_{-3}+\tfrac{3}{2}\,d_{-4}+\tfrac{7}{4}\,e_{-5}+\ldots.) \\
F^{\text{iv}}(t) &= \frac{1}{\omega^4}\,(d_{-4}+2e_{-5}+\ldots.) \\
F^{\text{v}}(t) &= \frac{1}{\omega^5}\,(e_{-5}+\ldots.) \\
&\ldots\ldots\ldots
\end{aligned}\right\} \quad (170)$$

As above, we emphasize the relative importance of the *first* derivative in practice: thus, for brevity, we write the first of equations (169) in the form

$$F^{\prime}(t-n\omega) = \frac{1}{\omega}\,(a_{-1}-B'b_{-2}+C'c_{-3}-D'd_{-4}+E'e_{-5}-\ldots.) \qquad (171)$$

the coefficients B', C', D', E' being taken from Table IV with the argument n.

Formulae (169), (170), and (171) are particularly useful in the computation of derivatives at or near the *end* of a series of functions.

Moreover, when the interval n approaches unity, formulae (169) and (171) are convenient for computing derivatives corresponding to the argument $t+n\omega$, since they enable us to proceed *backwards* from the argument $t+\omega$ with the interval $1-n$. We shall now solve several examples to illustrate these applications.

EXAMPLE I. — From the following ephemeris of the moon's right-ascension (a), compute the *hourly change* in a at the instant Feb. 3^d 20^h 24^m.

Date 1898 (d. h)	Moon's R.A. a (h m s)	Δ' (m s)	Δ'' (s)	Δ''' (s)	Δ^{iv} (s)	Δ^{v} (s)
Feb. 1 0	4 49 39.68					
		+26 21.18				
1 12	5 16 0.86		+ 4.81			
		26 25.99		−6.07		
2 0	5 42 26.85		− 1.26		+0.08	
		26 24.73		5.99		+0.54
2 12	6 8 51.58		7.25		0.62	
		26 17.48		5.37		+0.30
3 0	6 35 9.06		12.62		+0.92	
		26 4.86		−4.45		
3 12	7 1 13.92		−17.07			
		+25 47.79				
4 0	7 27 1.71					

Since the assigned unit of time is 1 hour, we have $\omega = 12$; hence, letting $t =$ Feb. 4^d 0^h, we find

$$n = \frac{4^d\ 0^h\ 0^m - 3^d\ 20^h\ 24^m}{12^h} = 0.30$$

which is the interval reckoned *backwards* from $t =$ Feb. 4^d 0^h. Denoting the quantity sought by Δa, we then have

$$\Delta a = F'(t-n\omega)$$

We therefore employ the formula (171): thus, taking the requisite differences from the given series, and their coefficients from Table IV, we obtain

			$a_{-1} =$	+25^m 47^s.79
$B' = -0.20$	$b_{-2} = -17^s.07$		$-B'b_{-2} =$	− 3.414
$C' = +0.07833$	$c_{-3} = -\ 4.45$		$+C'c_{-3} =$	− 0.349
$D' = -0.03800$	$d_{-4} = +\ 0.92$		$-D'd_{-4} =$	+ 0.035
$E' = +0.02009$	$e_{-5} = +\ 0.30$		$+E'e_{-5} =$	+ 0.006
			$\therefore\ \omega F'_{-n} =$	+25 44.07

Whence

$$\Delta a = F'_{-n} = 25^m\ 44^s.07 \div 12 = 2^m\ 8^s.672.$$

The change in a for *one minute* ($\Delta_1 a$) is simply

$$\Delta_1 a = \frac{\Delta a}{60} = 2^s.1445$$

EXAMPLE II. — From the preceding table of moon's R.A., compute the *hourly variation* in $\varDelta_1 a$ for Feb. $3^d\,12^h$; where, as above, $\varDelta_1 a$ denotes the change per *minute* in R.A.

Regarding one hour as the unit of time, it is clear that the value of $F''(t)$ given by (170) is sixty times the quantity sought: the expression for the required variation is therefore $\frac{1}{60}\,F''(t)$, where $t =$ Feb. $3^d\,12^h$. Accordingly, using the second of (170), we find

Hr. Var. in $\varDelta_1 a$, Feb. $3^d\,12^h$,

$$= \frac{1}{60} \times \frac{1}{(12)^2}\left(-12.62 - 5.37 + \tfrac{11}{12} \times 0.62 + \tfrac{5}{6} \times 0.54\right) = -0^s.00196$$

EXAMPLE III. — Given the following values of $F(T) \equiv \log_e T$: find $F'(T)$ for $T = 75$.

T	$F(T) \equiv \log_e T$	$\varDelta'$	$\varDelta''$	$\varDelta'''$	$\varDelta^{iv}$	$\varDelta^{v}$
45	3.80666					
		+10536				
50	3.91202		−1005			
		9531		+175		
55	4.00733		830		−41	
		8701		134		+ 9
60	4.09434		696		32	
		8005		102		+12
65	4.17439		594		−20	
		7411		+ 82		
70	4.24850		− 512			
		+ 6899				
75	4.31749					

Taking $t = 75$, and using the first of (170), we find

$$F'(t) = \frac{10^{-5}}{5}\left(6899 - \tfrac{512}{2} + \tfrac{82}{3} - \tfrac{20}{4} + \tfrac{12}{5}\right) = +0.01334$$

Since $F'(T) = \frac{1}{T}$, we observe that the *true mathematical value* of the computed quantity is—

$$F'(t) = \tfrac{1}{75} = +0.01333\tfrac{1}{3}$$

EXAMPLE IV. — From the preceding table of natural logarithms, compute $F''(T)$ for $T = 67$.

We let $t = 70$, and proceed by the second of (169), observing that

$$n = \frac{70 - 67}{5} = 0.60$$

Thus we obtain

$$b_{-2} = -0.00594$$

$$C'' = n - 1 = -0.40 \qquad c_{-3} = +102 \qquad -C''c_{-3} = + \quad 40.8$$

$$D'' = \tfrac{n^2}{2} - \tfrac{3}{2}n + \tfrac{11}{12} = +0.197 \qquad d_{-4} = -\ 32 \qquad +D''d_{-4} = - \quad 6.3$$

$$E'' = \tfrac{n^3}{6} - n^2 + \tfrac{7}{4}n - \tfrac{5}{6} = -0.107 \qquad e_{-5} = +\ 9 \qquad -E''e_{-5} = + \quad 1.0$$

$$\therefore\ \omega^2 F''_{-n} = -0.00558.5$$

$$\therefore\ F''_{-n} = -0.00022.3$$

The true value of this quantity is—

$$F''(T) = -\frac{1}{T^2} = -\frac{1}{(67)^2} = -0.00022.27\ .\ .\ .$$

61. *Derivatives from* STIRLING'S *Formula.* — When differences both preceding and following the function $F(t)$ are available, formulae more convenient and accurate than the foregoing may be employed. The most useful and important of these are derived from STIRLING'S Formula of interpolation (104), which is—

$$F_n = F_0 + na + Bb_0 + Cc + Dd_0 + Ee + \ .\ .\ .\ . \tag{172}$$

where the differences are taken according to the schedule on page 62, a, c, and e being the *mean* differences defined by (101); and where B, C, ... have the values

$$\left.\begin{aligned} B &= \frac{n^2}{2} \\ C &= \frac{n(n^2-1)}{6} = \frac{n^3}{6} - \frac{n}{6} \\ D &= \frac{n^2(n^2-1)}{24} = \frac{n^4}{24} - \frac{n^2}{24} \\ E &= \frac{n(n^2-1)(n^2-4)}{120} = \frac{n^5}{120} - \frac{n^3}{24} + \frac{n}{30} \\ &\ .\ .\ .\ .\ .\ .\ .\ .\ .\ .\ .\ .\ .\ .\ .\ . \end{aligned}\right\} \tag{173}$$

Whence, deriving the values of B', B'', ... , C', C'', ... , etc., from (173), and substituting these (with the above differences) in the general formulae (159), we get

$$\left.\begin{aligned} F'(t+n\omega) &= \frac{1}{\omega}\left(a + nb_0 + (\tfrac{n^2}{2} - \tfrac{1}{6})c + (\tfrac{n^3}{6} - \tfrac{n}{12})d_0 + (\tfrac{n^4}{24} - \tfrac{n^2}{8} + \tfrac{1}{30})e + \ .\ .\ .\ .\right) \\ F''(t+n\omega) &= \frac{1}{\omega^2}\left(b_0 + nc + (\tfrac{n^2}{2} - \tfrac{1}{12})d_0 + (\tfrac{n^3}{6} - \tfrac{n}{4})e + \ .\ .\ .\ .\right) \\ F'''(t+n\omega) &= \frac{1}{\omega^3}\left(c + nd_0 + (\tfrac{n^2}{2} - \tfrac{1}{4})e + \ .\ .\ .\ .\right) \\ F^{\text{iv}}(t+n\omega) &= \frac{1}{\omega^4}\left(d_0 + ne + \ .\ .\ .\ .\right) \\ F^{\text{v}}(t+n\omega) &= \frac{1}{\omega^5}\left(e + \ .\ .\ .\ .\right) \\ &\ .\ .\ .\ .\ .\ .\ .\ .\ .\ . \end{aligned}\right\} \tag{174}$$

Making $n = 0$ in (174), the latter become

$$\left.\begin{aligned}
F'(t) &= \frac{1}{\omega}\left(a - \tfrac{1}{6}c + \tfrac{1}{30}e - \ldots\ldots\right)\\
F''(t) &= \frac{1}{\omega^2}\left(b_0 - \tfrac{1}{12}d_0 + \ldots\ldots\right)\\
F'''(t) &= \frac{1}{\omega^3}\left(c - \tfrac{1}{4}e + \ldots\ldots\right)\\
F^{\text{iv}}(t) &= \frac{1}{\omega^4}\left(d_0 - \ldots\ldots\right)\\
F^{\text{v}}(t) &= \frac{1}{\omega^5}\left(e - \ldots\right)\\
&\ldots\ldots\ldots
\end{aligned}\right\} \tag{175}$$

Again, writing $-n$ for n in (174), we obtain

$$\left.\begin{aligned}
F'(t-n\omega) &= \frac{1}{\omega}\left(a - nb_0 + \left(\tfrac{n^2}{2} - \tfrac{1}{6}\right)c - \left(\tfrac{n^3}{6} - \tfrac{n}{12}\right)d_0 + \left(\tfrac{n^4}{24} - \tfrac{n^2}{8} + \tfrac{1}{30}\right)e - \ldots\ldots\right)\\
F''(t-n\omega) &= \frac{1}{\omega^2}\left(b_0 - nc + \left(\tfrac{n^2}{2} - \tfrac{1}{12}\right)d_0 - \left(\tfrac{n^3}{6} - \tfrac{n}{4}\right)e + \ldots\ldots\right)\\
F'''(t-n\omega) &= \frac{1}{\omega^3}\left(c - nd_0 + \left(\tfrac{n^2}{2} - \tfrac{1}{4}\right)e - \ldots\ldots\right)\\
F^{\text{iv}}(t-n\omega) &= \frac{1}{\omega^4}\left(d_0 - ne + \ldots\ldots\right)\\
F^{\text{v}}(t-n\omega) &= \frac{1}{\omega^5}\left(e - \ldots\ldots\right)
\end{aligned}\right\} \tag{176}$$

The coefficients for the computation of $F'(t \pm n\omega)$, namely—

$$\begin{aligned}
B' &= n\,, & D' &= \tfrac{n^3}{6} - \tfrac{n}{12}\\
C' &= \tfrac{n^2}{2} - \tfrac{1}{6}\,, & E' &= \tfrac{n^4}{24} - \tfrac{n^2}{8} + \tfrac{1}{30}
\end{aligned} \tag{177}$$

are given in Table V with the argument n. The quantity $F'(T)$ is thus readily computed (for any value of T) by either one or both of the formulae

$$F'(t+n\omega) = \frac{1}{\omega}\left(a + nb_0 + C'c + D'd_0 + E'e\right) \tag{178}$$

$$F'(t-n\omega) = \frac{1}{\omega}\left(a - nb_0 + C'c - D'd_0 + E'e\right) \tag{179}$$

in which the *odd* differences are algebraic *means* of the tabular differences, taken as indicated below :

T	$F(T)$	Δ'	Δ''	Δ'''	Δ^{iv}	Δ^{v}
$t-\omega$	F_{-1}		b'		d'	
		a'		c'		e'
t	F_0	(a)	b_0	(c)	d_0	(e)
		a_1		c_1		e_1
$t+\omega$	F_1		b_1		d_1	

The formulae (174) and (175) may also be obtained by the following method, which reverses the preceding order of development by deriving first the *particular*, and from the latter, the more *general* of the two groups in question.

Expanding $F(t+n\omega)$ by TAYLOR's Theorem, we have

$$F(t+n\omega) = F(t) + n\omega F'(t) + \frac{n^2\omega^2}{2!} F''(t) + \frac{n^3\omega^3}{3!} F'''(t) + \dots \qquad (180)$$

Arranging STIRLING's Formula (104) according to ascending powers of n, we find

$$\left.\begin{aligned} F(t+n\omega) = {} & F_0 + n\left(a - \tfrac{1}{6}c + \tfrac{1}{30}e - \dots\right) + \frac{n^2}{2!}\left(b_0 - \tfrac{1}{12}d_0 + \dots\right) \\ & + \frac{n^3}{3!}\left(c - \tfrac{1}{4}e + \dots\right) + \frac{n^4}{4!}\left(d_0 - \dots\right) + \frac{n^5}{5!}\left(e - \dots\right) \\ & + \dots\dots \end{aligned}\right\} \qquad (181)$$

Whence, by equating coefficients of like powers of n in the equivalent expressions (180) and (181), we obtain

$$\left.\begin{aligned} \omega F'(t) &= a - \tfrac{1}{6}c + \tfrac{1}{30}e - \dots & \omega^4 F^{iv}(t) &= d_0 - \dots \\ \omega^2 F''(t) &= b_0 - \tfrac{1}{12}d_0 + \dots & \omega^5 F^{v}(t) &= e - \dots \\ \omega^3 F'''(t) &= c - \tfrac{1}{4}e + \dots & & \dots\dots \end{aligned}\right\} \qquad (181a)$$

which agree with the formulae (175).

Again, by TAYLOR's Theorem, we have

$$\begin{aligned} F'(t+n\omega) &= F'(t) + n\omega F''(t) + \frac{n^2\omega^2}{2!} F'''(t) + \dots \\ F''(t+n\omega) &= F''(t) + n\omega F'''(t) + \frac{n^2\omega^2}{2!} F^{iv}(t) + \dots \\ & \dots\dots \end{aligned}$$

which may be written in the form

$$F'(t+n\omega) = \frac{1}{\omega}\left(\omega F'(t) + n\omega^2 F''(t) + \frac{n^2}{\lfloor 2}\,\omega^3 F'''(t) + \;.\;.\;.\;.\right)$$
$$F''(t+n\omega) = \frac{1}{\omega^2}\left(\omega^2 F''(t) + n\omega^3 F'''(t) + \frac{n^2}{\lfloor 2}\,\omega^4 F^{\text{iv}}(t) + \;.\;.\;.\;.\right)$$
$$.\;.$$

Substituting in these equations the expressions for $\omega F'(t)$, $\omega^2 F''(t)$, ... , as given by (181a), we get

$$\left.\begin{aligned}
F'(t+n\omega) &= \frac{1}{\omega}\Big[(a-\tfrac{1}{6}c+\tfrac{1}{30}e-\;.\;.\;)+n\,(b_0-\tfrac{1}{12}d_0+\;.\;.\;) \\
&\qquad +\frac{n^2}{\lfloor 2}(c-\tfrac{1}{4}e+\;.\;.\;)+\frac{n^3}{\lfloor 3}(d_0-\;.\;.\;)+\frac{n^4}{\lfloor 4}(e-\;.\;.\;)+\;.\;.\;\Big] \\
F''(t+n\omega) &= \frac{1}{\omega^2}\Big[(b_0-\tfrac{1}{12}d_0+\;.\;.\;)+n\,(c-\tfrac{1}{4}e+\;.\;.\;)+\frac{n^2}{\lfloor 2}(d_0-\;.\;.\;) \\
&\qquad +\frac{n^3}{\lfloor 3}(e-\;.\;.\;)+\;.\;.\;\Big] \\
F'''(t+n\omega) &= \frac{1}{\omega^3}\Big[(c-\tfrac{1}{4}e+\;.\;.\;)+n\,(d_0-\;.\;.\;)+\frac{n^2}{\lfloor 2}(e-\;.\;.\;)+\;.\;.\;\Big] \\
F^{\text{iv}}(t+n\omega) &= \frac{1}{\omega^4}\Big[(d_0-\;.\;.\;)+n\,(e-\;.\;.\;)+\;.\;.\;\Big] \\
F^{\text{v}}(t+n\omega) &= \frac{1}{\omega^5}\Big[(e-\;.\;.\;)+\;.\;.\;\Big] \\
.\;.\;.\;.\;.\;.\;.\;.\;.\;.\;.\;.\;.
\end{aligned}\right\} \quad (182)$$

These expressions, upon being arranged according to the successive orders of differences, will be found identical with the formulae (174). For some purposes, however, the present form is more convenient.

It is quite common, particularly in an astronomical ephemeris, to tabulate the values of $F'(T)$ corresponding to the tabular values of $F(T)$. Such a table would run as follows:*

T	$F(T)$	$F'(T)$
$t-2\omega$	F_{-2}	$F'(t-2\omega)$
$t-\omega$	F_{-1}	$F'(t-\omega)$
t	F_0	$F'(t)$
$t+\omega$	F_1	$F'(t+\omega)$
$t+2\omega$	F_2	$F'(t+2\omega)$

* It is evident that $F'(t+n\omega)$ can be derived from the column $F'(T)$ by direct interpolation: moreover, when the tabular values of $F'(T)$ are thus available, this method of computing $F'(t+n\omega)$ is more expeditious than the use of formula (178).

The first of the formulae (175) is almost invariably used for this purpose, because of its simplicity and rapid convergence; this formula is, in fact, the most important and useful of those which pertain to the computation of derivatives. For this reason we formulate the following

RULE for computing the first derivative of a tabular function corresponding to one of the given functional values: *From the mean of the two first differences which immediately precede and follow the function in question, subtract one-sixth* ($\frac{1}{6}$) *the mean of the corresponding third differences, and divide the result by the tabular interval.* This rule neglects only 5th and higher differences. To include 5th and 6th differences, *add to the above terms (before dividing by ω) one-thirtieth* ($\frac{1}{30}$) *the mean of the corresponding fifth differences, and divide by ω as before.*

It will evidently suffice, in most cases, to apply only the first part of the above rule.

Several examples will now be solved as an exercise in the use of the preceding formulae.

EXAMPLE I. — Given the following ephemeris of the sun's declination (δ): compute the *hourly difference* in δ for the dates Jan. 7, 10, 13, and 16.

Date 1898	Sun's Decl. δ	Δ'	Δ''	Δ'''	a	$-\frac{1}{6}c$	Diff. for 1 hour
	° ′ ″	′ ″	′ ″	″	″	″	″
Jan. 1	−22 59 2.4						
		+17 23.9					
4	22 41 38.5		+4 2.2				
		21 26.1		−5.3			
7	22 20 12.4		3 56.9		+1404.55	+0.98	+19.52
		25 23.0		6.4			
10	21 54 49.4		3 50.5		1638.25	1.12	22.77
		29 13.5		7.1			
13	21 25 35.9		3 43.4		1865.20	1.27	25.92
		32 56.9		8.1			
16	20 52 39.0		3 35.3		+2084.55	+1.45	+28.97
		36 32.2		−9.3			
19	20 16 6.8		+3 26.0				
		+39 58.2					
22	−19 36 8.6						

The term $\frac{1}{30}e$ in the first of (175) is here insensible; hence, for each of the given dates we have only to compute the quantity

$$F'(t) = \frac{1}{\omega}(a - \tfrac{1}{6}c)$$

Accordingly, in column a we write the required *mean* first differences, expressed in seconds of arc. The next column contains *minus* one-sixth of the corresponding mean third differences. Finally, since $\omega = 72$ hours, we write in the last column $\frac{1}{72}$ of the quantities formed by summing the corresponding terms of the two preceding columns. We thus obtain the hourly differences required.

EXAMPLE II.—Compute, from the ephemeris of the last example, the *daily motion* in declination for the date Jan. $6^d\ 13^h\ 30^m$.

We proceed *backwards* from Jan. 7, using the formula (179), and taking the coefficients from Table V with the argument

$$n = \frac{7^d\ 0^h\ 0^m - 6^d\ 13^h\ 30^m}{3^d} = \frac{10^h.5}{72^h} = 0.14583$$

Thus we find

				a	$= +23'\ 24''.55$
$n =$	0.14583	$b_0 =$	$+236''.9$	$-nb_0$	$= -\quad 34.55$
$C' =$	-0.1560	$c =$	$-\quad 5.85$	$+C'c$	$= +\quad 0.91$
$D' =$	-0.012	$d_0 =$	$-\quad 1.1$	$-D'd_0$	$= -\quad 0.01$
				$\therefore\ \omega F'_{-n}$	$= +22\ 50.90$

Whence, for the daily motion in δ, Jan. $6^d\ 13^h\ 30^m$, we obtain

$$F'_{-n} = 22'\ 50''.90 \div 3 = +7'\ 36''.97$$

EXAMPLE III.—The following table gives $F(T) \equiv e^T$, where e denotes the base of natural logarithms: compute $F'(T)$ for $T = 0.30$.

T	$F(T) \equiv e^T$	Δ'	Δ''	Δ'''	Δ^{iv}	Δ^{v}
0.0	1.000000					
		+105171				
0.1	1.105171		+11061			
		116232		+1163		
0.2	1.221403		12224		+123	
		128456		1286		+11
0.3	1.349859		13510		134	
		141966		1420		18
0.4	1.491825		14930		152	
		156896		1572		+10
0.5	1.648721		16502		+162	
		173398		+1734		
0.6	1.822119		+18236			
		+191634				
0.7	2.013753					

Using the first of (175), we find

$$F'(0.30) = \frac{10^{-6}}{0.1}\left(135211 - \tfrac{1353}{6} + \tfrac{14.5}{30}\right) = 1.34986$$

It will be observed that our result is substantially equal to the value of $F(T)$ for the same argument, $T = 0.30$: this is required by the relation

$$F(T) = F'(T) = F''(T) = \ldots\ldots = e^T$$

EXAMPLE IV.—From the table of Example III, compute $F''(T)$ for $T = 0.462$.

Taking $t = 0.4$ and $n = 0.62$, we obtain, by means of the second of (174),

		b_0	$= +0.014930$
$n = 0.62$	$c = +1496$	nc	$= +\quad 927.5$
$D'' = \frac{n^2}{2} - \frac{1}{12} = +0.1089$	$d_0 = +\ 152$	$D''d_0$	$= +\quad 16.6$
$E'' = \frac{n^3}{6} - \frac{n}{4} = -0.115$	$e = +\ 14$	$E''e$	$= -\quad 1.6$
		$\therefore\ \omega^2 F_n''$	$= +0.0158725$
		$\therefore\ F_n''$	$= +1.58725$

The true mathematical value is—

$$F''(T) = F(T) = e^T = e^{0.462} = 1.587245\ \ldots$$

62. *Derivatives from* BESSEL'S *Formula.*—Other useful formulae, convenient for the computation of tabular derivatives, are those derived from BESSEL'S Formula of interpolation (111). The latter may be written in the form

$$F_n = F_0 + na_1 + Bb + Cc_1 + Dd + Ee_1 + \ldots . \tag{183}$$

where the differences are taken as in the schedule on page 62, b and d being the *mean* differences defined by (106); and where B, C, ... have the following values :

$$\left.\begin{aligned}
B &= \frac{n(n-1)}{2} = \frac{n^2}{2} - \frac{n}{2} \\
C &= \frac{n(n-1)(n-\frac{1}{2})}{6} = \frac{n^3}{6} - \frac{n^2}{4} + \frac{n}{12} \\
D &= \frac{(n+1)n(n-1)(n-2)}{24} = \frac{n^4}{24} - \frac{n^3}{12} - \frac{n^2}{24} + \frac{n}{12} \\
E &= \frac{(n+1)n(n-1)(n-2)(n-\frac{1}{2})}{120} = \frac{n^5}{120} - \frac{n^4}{48} + \frac{n^3}{48} - \frac{n}{120} \\
&\ldots\ldots\ldots\ldots\ldots\ldots
\end{aligned}\right\} \tag{184}$$

Deriving from (184) the values of B', B'', ... , C', C'', ... , etc., according to (154), and substituting these in the general formulae (159), we obtain

$$\left.\begin{aligned}
F'\ (t+n\omega) &= \frac{1}{\omega}\Big(a_1+(n-\tfrac{1}{2})b+(\tfrac{n^2}{2}-\tfrac{n}{2}+\tfrac{1}{12})c_1+(\tfrac{n^3}{6}-\tfrac{n^2}{4}-\tfrac{n}{12}+\tfrac{1}{12})d \\
&\qquad +(\tfrac{n^4}{24}-\tfrac{n^3}{12}+\tfrac{n}{24}-\tfrac{1}{120})e_1+\ .\ .\ .\ .\Big) \\
F''\ (t+n\omega) &= \frac{1}{\omega^2}\Big(b+(n-\tfrac{1}{2})c_1+(\tfrac{n^2}{2}-\tfrac{n}{2}-\tfrac{1}{12})d+(\tfrac{n^3}{6}-\tfrac{n^2}{4}+\tfrac{1}{24})e_1+\ .\ .\ .\ .\Big) \\
F'''\ (t+n\omega) &= \frac{1}{\omega^3}\Big(c_1+(n-\tfrac{1}{2})\,d+(\tfrac{n^2}{2}-\tfrac{n}{2})\,e_1+\ .\ .\ .\ .\Big) \\
F^{iv}\ (t+n\omega) &= \frac{1}{\omega^4}\Big(d+(n-\tfrac{1}{2})\,e_1+\ .\ .\ .\ .\Big) \\
F^{v}\ (t+n\omega) &= \frac{1}{\omega^5}\Big(e_1+\ .\ .\ .\ .\Big) \\
.\ .\ .\ .\ .\ .\ .\ .\ .\ .\ .\ .
\end{aligned}\right\}\qquad(185)$$

Putting $n=0$ in (185), we get

$$\left.\begin{aligned}
F'\ (t) &= \frac{1}{\omega}\ (a_1-\tfrac{1}{2}b+\tfrac{1}{12}c_1+\tfrac{1}{12}d-\tfrac{1}{120}e_1-\ .\ .\ .\ .\ ) \\
F''\ (t) &= \frac{1}{\omega^2}\ (b-\tfrac{1}{2}c_1-\tfrac{1}{12}d+\tfrac{1}{24}e_1+\ .\ .\ .\ .\ ) \\
F'''\ (t) &= \frac{1}{\omega^3}\ (c_1-\tfrac{1}{2}d+0^*+\ .\ .\ .\ .\ ) \\
F^{iv}\ (t) &= \frac{1}{\omega^4}\ (d-\tfrac{1}{2}e_1-\ .\ .\ .\ .\ ) \\
F^{v}\ (t) &= \frac{1}{\omega^5}\ (e_1-\ .\ .\ .\ .\ ) \\
.\ .\ .\ .\ .\ .\ .\ .\ .\ .
\end{aligned}\right\}\qquad(186)$$

Again, putting $n=\frac{1}{2}$ in (185), we obtain the following simple formulae:

$$\left.\begin{aligned}
F'\ (t+\tfrac{1}{2}\omega) &= \frac{1}{\omega}\ (a_1-\tfrac{1}{24}c_1+\tfrac{3}{640}e_1-\ .\ .\ .\ .\ ) \\
F''\ (t+\tfrac{1}{2}\omega) &= \frac{1}{\omega^2}\ (b-\tfrac{5}{24}d+\ .\ .\ .\ .\ ) \\
F'''\ (t+\tfrac{1}{2}\omega) &= \frac{1}{\omega^3}\ (c_1-\tfrac{1}{8}e_1+\ .\ .\ .\ .\ ) \\
F^{iv}\ (t+\tfrac{1}{2}\omega) &= \frac{1}{\omega^4}\ (d-\ .\ .\ .\ .\ ) \\
F^{v}\ (t+\tfrac{1}{2}\omega) &= \frac{1}{\omega^5}\ (e_1-\ .\ .\ .\ .\ )\ .
\end{aligned}\right\}\qquad(187)$$

which determine the derivatives of $F(T)$ at points *midway* between the tabular values of the function. It is important to observe that,

* The coefficient of e_1 vanishes.

unless third differences are considerable, a close *approximation* to $F'(t+\frac{1}{2}\omega)$ is given by the simple expression

$$F'(t+\tfrac{1}{2}\omega) = \frac{a_1}{\omega} = \frac{F_1 - F_0}{\omega} \qquad (187a)$$

which differs from the *exact* formula only by the omission of the small quantity

$$\frac{1}{\omega}\left(-\tfrac{1}{24}A''' + \ldots\ldots\right)$$

The formulae for the derivatives of $F(t-n\omega)$ are deduced from (111a). Let us put, for brevity,

$$\bar{b} = \tfrac{1}{2}(b_0 + b') \quad , \quad \bar{d} = \tfrac{1}{2}(d_0 + d') \qquad (188)$$

and (111a) becomes

$$F_{-n} = F_0 - na' + B\bar{b} - Cc' + D\bar{d} - Ee' + \ldots\ldots \qquad (189)$$

Comparing this expression with the general formula (160), we find that $\alpha, \beta, \gamma, \delta, \epsilon, \ldots\ldots$, in the latter, are replaced by $a', \bar{b}, c', \bar{d}, e', \ldots\ldots$ in (189); hence, observing these changes, and substituting the above determined values of $B', B'', \ldots\ldots, C', C'', \ldots\ldots$, etc., in the formulae (161), we obtain

$$\left.\begin{aligned}
F'(t-n\omega) &= \frac{1}{\omega}\Big(a' - (n-\tfrac{1}{2})\bar{b} + (\tfrac{n^2}{2} - \tfrac{n}{2} + \tfrac{1}{12})c' - (\tfrac{n^3}{6} - \tfrac{n^2}{4} - \tfrac{n}{12} + \tfrac{1}{12})\bar{d} \\
&\qquad + (\tfrac{n^4}{24} - \tfrac{n^3}{12} + \tfrac{n}{24} - \tfrac{1}{120})e' - \ldots\ldots\Big) \\
F''(t-n\omega) &= \frac{1}{\omega^2}\Big(\bar{b} - (n-\tfrac{1}{2})c' + (\tfrac{n^2}{2} - \tfrac{n}{2} - \tfrac{1}{12})\bar{d} - (\tfrac{n^3}{6} - \tfrac{n^2}{4} + \tfrac{1}{24})e' + \ldots\ldots\Big) \\
F'''(t-n\omega) &= \frac{1}{\omega^3}\Big(c' - (n-\tfrac{1}{2})\bar{d} + (\tfrac{n^2}{2} - \tfrac{n}{2})e' - \ldots\ldots\Big) \\
F^{\text{iv}}(t-n\omega) &= \frac{1}{\omega^4}\Big(\bar{d} - (n-\tfrac{1}{2})e' + \ldots\ldots\Big) \\
F^{\text{v}}(t-n\omega) &= \frac{1}{\omega^5}\Big(e' - \ldots\ldots\Big) \\
&\ldots\ldots\ldots\ldots
\end{aligned}\right\} \qquad (190)$$

The values of B', C', D', and E', as computed from the expressions

$$\left.\begin{aligned}
B' &= n - \tfrac{1}{2} \quad , & D' &= \tfrac{n^3}{6} - \tfrac{n^2}{4} - \tfrac{n}{12} + \tfrac{1}{12} \\
C' &= \tfrac{n^2}{2} - \tfrac{n}{2} + \tfrac{1}{12} \quad , & E' &= \tfrac{n^4}{24} - \tfrac{n^3}{12} + \tfrac{n}{24} - \tfrac{1}{120}
\end{aligned}\right\} \qquad (191)$$

are given in Table VI with the argument n. By means of these coefficients, values of $F'(T)$ are readily computed from either one of the formulae

$$F'(t+n\omega) = \frac{1}{\omega}(a_1 + B'b + C'c_1 + D'd + E'e_1) \tag{192}$$

$$F'(t-n\omega) = \frac{1}{\omega}(a' - B'\bar{b} + C'c' - D'\bar{d} + E'e') \tag{193}$$

in which the *even* differences are *means*, taken as indicated below:

T	$F(T)$	Δ'	Δ''	Δ'''	Δ^{iv}	Δ^{v}
$t-\omega$	F_{-1}		b'		d'	
		a'	$(\bar{b})$	c'	$(\bar{d})$	e'
t	F_0		b_0		d_0	
		a_1	(b)	c_1	(d)	e_1
$t+\omega$	F_1		b_1		d_1	

Several examples will now be solved.

EXAMPLE I. — Given the following table of natural sines:

T	$F(T) \equiv \sin T$	Δ'	Δ''	Δ'''	Δ^{iv}
40°	0.6427876				
		+263430			
42	0.6691306		−8152		
		255278		−312	
44	0.6946584		8464		+12
		246814		300	
46	0.7193398		8764		+10
		238050		−290	
48	0.7431448		−9054		
		+228996			
50	0.7660444				

Let it be required to find $F'(T)$ for $T = 45°$.

Taking $t = 44°$, we have

$$\omega = 2° = \frac{\pi}{90} = 0.0349066 \qquad n = \tfrac{1}{2}$$

Hence, using the first of (187), we find

$$\begin{aligned} & & a_1 &= +0.0246814 \\ c_1 &= -300\,; & -\tfrac{1}{24}c_1 &= +\qquad 12.5 \\ \hline & & \therefore\ \omega F'_{\frac{1}{2}} &= +0.0246826.5 \\ & & \therefore\ F'_{\frac{1}{2}} &= +0.707106 \end{aligned}$$

The true value of this quantity is —

$$F'(T) = \cos T = \cos 45° = 0.707107$$

EXAMPLE II. — From the preceding table, compute the value of $F''(T)$ for $T = 44° 48'$.

We take $t = 44°$; hence $n = 0.40$. Accordingly, from the second of (185), we obtain

$$
\begin{array}{llr}
 & & b = -0.0008614 \\
C'' = n - \frac{1}{2} = -0.10 & c_1 = -300 & C''c_1 = + \qquad 30 \\
D'' = \frac{n^2}{2} - \frac{n}{2} - \frac{1}{12} = -0.203 & d = +\ 11 & D''d = - \qquad 2 \\
\hline
 & & \therefore\ \omega^2 F''_n = -0.0008586 \\
 & & \therefore\ F''_n = -0.70465
\end{array}
$$

The actual value is —

$$F''(T) = -\sin T = -\sin 44° 48' = -0.70463$$

EXAMPLE III. — The table below gives the Washington mean time of moon's upper transit at the meridian of Washington:

WASHINGTON MOON CULMINATIONS.

Date 1898	Mean Time of Transit	Δ'	Δ''	Δ'''	Δ^{iv}
	h m	m	m	m	m
Mar. 22	0 15.57				
		+45.43			
23	1 1.00		+0.86		
		46.29		+0.44	
24	1 47.29		1.30		−0.38
		47.59		+0.06	
25	2 34.88		1.36		0.36
		48.95		−0.30	
26	3 23.83		1.06		−0.37
		50.01		−0.67	
27	4 13.84		+0.39		
		+50.40			
28	5 4.24				

Before proposing an example from this ephemeris, it is proper to remark that the tabular function is the *time* of the moon's arrival at a succession of meridians (in reality one fixed meridian) whose common difference of longitude is 24 hours. The *argument* of the series is therefore the terrestrial *longitude* traversed by the moon, counted west from the Washington meridian: the *interval* of this argument is 24 hours of longitude.

Now, let D denote the *difference in time of transit for 1 hour of longitude.* This quantity is simply the first derivative of the tabular function: computed for the instant of transit at a meridian l hours west of Washington, the quantity D expresses the amount by which the local time of transit at the meridian $l + 1$ hours would exceed the local time of transit at the meridian l hours, supposing the rate of retardation to remain constant between the two transits, and equal to what it is at the moment of the first. Thus, if D_0 is the value of D for the instant of transit at Washington on Mar. 24, the local time of moon's transit at a station 20 minutes west of Washington is given with sufficient precision by the formula

$$\tau = \text{Mar. } 24^{\text{d}}\ 1^{\text{h}}\ 47^{\text{m}}.29 + \tfrac{1}{3} D_0$$

Now, by the first of equations (186), we find for the value of D_0,

$$D_0 = F'(t) = \tfrac{1}{24}\left(47.59 - \tfrac{1.33}{2} + \tfrac{0.06}{12} - \tfrac{0.37}{12}\right) = 1^{\text{m}}.954$$

Hence the preceding equation gives

$$\tau = \text{Mar. } 24^{\text{d}}\ 1^{\text{h}}\ 47^{\text{m}}.94$$

In this manner the local time of transit is simply and accurately determined for any number of stations within half an hour of the Washington meridian.

To find the local time of moon's transit over a meridian 3 hours west of Washington, on the 24th day of March, we have only to interpolate the Washington time of transit between the tabular values for Mar. 24 and Mar. 25, as given above, the interval from the former being

$$n = \frac{3^{\text{h}}}{24^{\text{h}}} = 0.125$$

Finally, if it were required to compute the local time of transit for *several* stations whose longitudes range from $2\frac{1}{2}$ to $3\frac{1}{2}$ hours west of Washington, we should find the time for the 3 hour meridian by direct interpolation, as explained above. We should also compute $D = F'(T)$ for the same meridian; that is, for $n = 0.125$. Then the local time of transit at any adjacent meridian, whose longitude from Washington is $3^{\text{hr}} + \lambda^{\text{min}}$, is given by the simple formula

$$\tau = \tau_1 + \frac{\lambda}{60} D$$

where τ_1 is the time of transit at the 3 hour meridian.

EXAMPLE IV — From the preceding ephemeris, compute the *difference in time of transit for 1 hour of longitude* (D) at the instant of moon's transit over the meridian of San Francisco, Mar. 25, 1898; the longitude from Washington being taken as $3^h\ 1^m\ 30^s = 3^h.025$.

Here we use the formula (192): thus, taking the coefficients from Table VI (with the argument $n = 3.025 \div 24 = 0.12604$), and the differences from the given ephemeris, we obtain

$$\begin{array}{llllll} & & & & a_1 & = +48^m.95 \\ B' & = -0.3740 & b & = +1.21 & B'b & = -\ 0.453 \\ C' & = +0.0282 & c_1 & = -0.30 & C'c_1 & = -\ 0.008 \\ D' & = +0.0692 & d & = -0.365 & D'd & = -\ 0.025 \\ \hline & & & & \therefore \omega F'_n & = +48.464 \end{array}$$

$$\therefore\ D = F'_n = 48^m.464 \div 24 = +2^m.019$$

EXAMPLE V. — Use the above table of Moon Culminations to find the variation in D for 24 hours of longitude, at the instant of moon's *lower* transit over the meridian of Washington, Mar. 24, 1898.

The *lower* transit at Washington is evidently the *upper* transit over the meridian 12 hours west. Hence, denoting the required variation by V, and regarding 1 hour of longitude as the unit, we find by the second of (187), for $t =$ Mar. 24,

$$V = 24F''(t + \tfrac{1}{2}\omega) = \frac{24}{\omega^2}\,(b - \tfrac{5}{24}\,d + \ .\ .\ .\ .\ )$$

$$= \tfrac{1}{24}\,(1.33 + \tfrac{5}{24} \times 0.37) = +0^m.059$$

63. *Interpolation of Functions by Means of their Tabular First Derivatives.* — As already observed, it frequently happens that a table giving $F(T)$ also contains the values of $F'(T)$ which correspond to the tabular functions. The object in thus tabulating the derivative is to facilitate the interpolation of intermediate values of $F(T)$. To derive the formula upon which this method is based, we consider the schedule below, where the differences are those of the series $F'(T)$:

T	$F(T)$	$F'(T)$	1st Diff.	2d	3d
$t - 2\omega$	F_{-2}	F'_{-2}			
			a''		
$t - \omega$	F_{-1}	F'_{-1}		β'	
			a'		γ'
t	F_0	F'_0		β_0	
			a_1		γ_1
$t + \omega$	F_1	F'_1		β_1	
			a_2		
$t + 2\omega$	F_2	F'_2			

We shall assume that the differences of $F(T)$ beyond Δ^{iv} may be disregarded; hence the differences of $F'(T)$ beyond γ may be neglected in the above schedule. Now, by TAYLOR'S Theorem, we have

$$F_n = F_0 + n\omega F_0' + \frac{n^2\omega^2}{\underline{|2}} F_0'' + \frac{n^3\omega^3}{\underline{|3}} F_0''' + \frac{n^4\omega^4}{\underline{|4}} F_0^{iv} + \ldots \qquad (194)$$

Again, since

$$F_0'' = \frac{dF'}{dt}, \qquad F_0''' = \frac{d^2F'}{dt^2}, \qquad F_0^{iv} = \frac{d^3F'}{dt^3}, \qquad \text{etc.},$$

we obtain, by means of the formulae (175),

$$F_0'' = \frac{1}{\omega}(\alpha - \tfrac{1}{6}\gamma) \quad , \quad F_0''' = \frac{\beta_0}{\omega^2} \quad , \quad F_0^{iv} = \frac{\gamma}{\omega^3} \qquad (195)$$

in which we have put, for brevity,

$$\alpha = \tfrac{1}{2}(\alpha' + \alpha_1) \quad , \quad \gamma = \tfrac{1}{2}(\gamma' + \gamma_1) \qquad (196)$$

Substituting these expressions for F_0'', F_0''', and F_0^{iv} in (194), the latter becomes

$$F_n = F_0 + n\omega F_0' + \frac{n^2\omega}{\underline{|2}}(\alpha - \tfrac{1}{6}\gamma) + \frac{n^3\omega}{\underline{|3}}\beta_0 + \frac{n^4\omega}{\underline{|4}}\gamma,$$

which may be written

$$F_n = F_0 + n\omega\left(F_0' + \tfrac{n}{2}\alpha + \tfrac{n^2}{6}\beta_0 + \tfrac{n}{12}(\tfrac{n^2}{2} - 1)\gamma\right) \qquad (197)$$

By means of this formula we compute F_n in terms of the differences of $F'(T)$, instead of the differences of $F(T)$ direct, as in the usual formulae of interpolation.

Substituting $-n$ for n in (197), we have

$$F_{-n} = F_0 - n\omega\left(F_0' - \tfrac{n}{2}\alpha + \tfrac{n^2}{6}\beta_0 - \tfrac{n}{12}(\tfrac{n^2}{2} - 1)\gamma\right) \qquad (198)$$

The values of

$$\mathrm{B} \equiv \tfrac{n^2}{6} \quad , \quad \Gamma \equiv \tfrac{n}{12}(\tfrac{n^2}{2} - 1) \qquad (199)$$

are given in Table VIII with the argument n. By means of these coefficients we readily compute

$$F_n = F_0 + n\omega\,(F_0' + \tfrac{n}{2}\,\alpha + \mathrm{B}\beta_0 + \Gamma\gamma) \qquad (200)$$

$$F_{-n} = F_0 - n\omega\,(F_0' - \tfrac{n}{2}\,\alpha + \mathrm{B}\beta_0 - \Gamma\gamma) \qquad (201)$$

The coefficients in Table VIII are not extended beyond $n = 0.60$, since by this method it is invariably more convenient to proceed from the *nearest* function F_0.

EXAMPLE.—From the *American Ephemeris* for 1898 we take the heliocentric longitude of *Mercury*, together with the *daily motion* in longitude, for a portion of the month of October. The differences of the daily motion are then taken, as shown below:

Date 1898	Helioc. Long. of *Mercury*	Daily Motion	α	β	γ	δ
	° ′ ″	° ′ ″	′ ″	′ ″	″	″
Oct. 11	176 51 7.8	4 2 34.3	−14 0.0			
13	184 41 59.2	3 48 34.3	12 17.5	+1 42.5	−5.4	
15	192 6 33.3	3 36 16.8	10 40.4	1 37.1	6.1	−0.7
17	199 8 10.6	3 25 36.4	9 9.4	1 31.0	−7.4	−1.3
19	205 49 59.6	3 16 27.0	− 7 45.8	+1 23.6		
21	212 14 54.7	3 8 41.2				

Let it be required to find the heliocentric longitude of *Mercury* for the date Oct. 15^d 14^h 24^m.0.

Here we have

$$t = \text{Oct. } 15^d \qquad T = \text{Oct. } 15^d\, 14^h\, 24^m.0 = \text{Oct. } 15^d.60$$

$$\omega = 2^d \qquad n\omega = T - t = 0^d.60 \qquad n = 0.30$$

Hence, using Table VIII, in connection with (200), we obtain

$F_0 = 192°\ 6'\ 33''.3$		$F_0' = +3°\ 36'\ 16''.8$
$\frac{n}{2} = +0.15$	$\alpha = -11'\ 28''.95$	$\frac{n}{2}\alpha = -\quad 1\ 43.34$
$\mathrm{B} = +0.0150$	$\beta_0 = +\ 1\ 37.1$	$\mathrm{B}\beta_0 = +\quad 1.46$
$\Gamma = -0.0239$	$\gamma = -\ 0\ \ 5.75$	$\Gamma\gamma = +\quad 0.14$
		$\therefore$ Sum, $D = +3\ 34\ 35.06$

Whence

$$F_n = F_0 + n\omega\,.\,D = 194°\ 15'\ 18''.3$$

Differencing the given series of longitudes and applying BESSEL's Formula of interpolation, we find

$$F_n = 194° \; 15' \; 18''.2$$

64. *Application of the Preceding Method of Interpolation when the Second Differences of the Series $F(T)$ are Nearly Constant.*— When the 3d and 4th differences of $F(T)$ are small enough to be neglected, we may omit the terms containing β_0 and γ in the formulae (197) and (198): we therefore obtain

$$F_n = F_0 + n\omega\,(F_0' + \tfrac{n}{2}\,\alpha) \qquad (202)$$

$$F_{-n} = F_0 - n\omega\,(F_0' - \tfrac{n}{2}\,\alpha) \qquad (203)$$

It will be interesting to determine the error of these approximate formulae as applied when the 3d differences of $F(T)$ are appreciable. For this purpose we write (197) in the form

$$F_n = F_0 + n\omega\,(F_0' + \tfrac{n}{2}\,\alpha) + \tfrac{n^3}{6}\,\omega\beta_0 + (\tfrac{n^4}{24} - \tfrac{n^2}{12})\,\omega\gamma$$

Hence, if we disregard 4th differences of $F(T)$, and thus neglect γ, it follows that the error in question is—

$$\epsilon = \pm \tfrac{n^3}{6}\,\omega\beta_0 \qquad (204)$$

Now, from (175), we have

$$F'''(t) = \frac{c}{\omega^3} = \frac{\Delta'''}{\omega^3}$$

also, from (195),

$$F'''(t) = \frac{\beta_0}{\omega^2}$$

Whence

$$\omega\beta_0 = c = \Delta''' \qquad (205)$$

and (204) becomes

$$\epsilon = \pm \tfrac{n^3}{6}\,\Delta''' \qquad (206)$$

Since in practice the maximum value of n is 0.50, it follows that the maximum error resulting from an application of the formulae (202) and (203), when 3d differences of $F(T)$ are sensible, is $\frac{1}{48}\Delta'''$. Hence, even when third differences are considerable, these formulae are sufficiently accurate for many purposes.

That the formulae (202) and (203) are *rigorously* true when the 3d differences of $F(T)$ are *zero* may be clearly shown from geometrical considerations, as follows:

The 2d differences of $F(T)$ being supposed constant, it follows from Theorem VI that the function is necessarily of the form

$$F(T) \equiv a_0T^2 + a_1T + a_2 \tag{207}$$

Now, if in the accompanying figure we draw the rectangular coördinate axes OT and OY, and plot the curve defined analytically by (207) (regarding $y = F(T)$ as the ordinate corresponding to the abscissa T), it is evident that we obtain a *parabola* whose axis is parallel to OY.

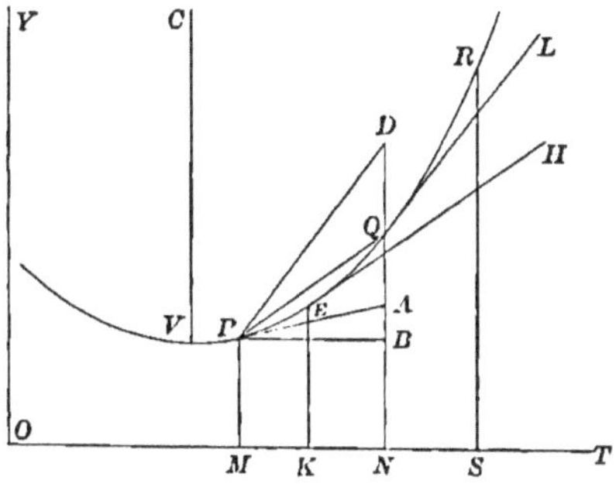

Let us now take

$$OM = t$$
$$OS = t + \omega$$
$$ON = t + n\omega$$

Whence

$$MN = n\omega$$
$$MP = F(t) = F_0$$
$$NQ = F(t+n\omega) = F_n$$

Draw the tangents PA, QL; also, draw $PD \parallel QL$ and $PB \parallel MN$. Then, denoting $\frac{dF}{dT}$ by F_n', we have

$$F_0' = \tan APB$$
$$F_n' = \tan DPB$$

Hence we find

$$NA = MP + PB \tan APB = F_0 + n\omega F_0'$$
$$ND = MP + PB \tan DPB = F_0 + n\omega F_n'$$

It is therefore evident that to find $NQ = F_n$, which lies between NA and ND, we must employ a value of F' somewhere between the values F_0' and F_n'. Now, let KE be the ordinate erected at the middle point of MN, and EH the tangent at E. Then, by an elementary theorem of the parabola, the chord PQ is *parallel* to EH, and we have, therefore,

$$F_n = NQ = MP + PB \tan QPB = F_0 + n\omega F'_{\frac{n}{2}} \qquad (208)$$

which agrees with the formula (202).

We have shown above that the maximum error produced by applying this formula when the second differences of $F(T)$ are *not* constant, is $\frac{1}{48}\Delta'''$. Hence, unless the 2d differences of $F'(T)$ are considerable, we may compute F_n by the following

RULE: *Find by simple interpolation the value of the tabular derivative which belongs midway between the required function and the nearest tabular function (F_0); multiply this quantity ($F'_{\frac{n}{2}}$) by the units contained in the entire interval ($T—t$), and apply the product to F_0.*

EXAMPLE I.—Given the following ephemeris of the moon's declination (δ): compute the value for the date July 9[d] 5[h] 18[m].0.

Date 1898	Moon's Decl. δ	Diff. for 1 Minute	α	β
d h	° ′ ″	″	″	″
July 9 1	+6 2 14.1	+13.876	−0.144	
9 4	6 43 39.0	13.732	0.150	−.006
9 7	7 24 37.4	13.582	−0.160	−.010
July 9 10	+8 5 8.0	+13.422		

Here $\omega = 3^h = 180^m$; hence, taking $t =$ July 9[d] 4[h], we find

$$n = \frac{78^m}{180^m} = 0.433 \qquad \frac{n}{2} = 0.217$$

Accordingly, the value of F' interpolated for *half* the interval, or 39 minutes, is—

$$F'_{\frac{n}{2}} = F'_0 + \tfrac{n}{2}\alpha = 13''.732 - 0.217 \times 0''.147 = 13''.700$$

Whence we obtain

$$\delta = 6^\circ\ 43'\ 39''.0 + 78 \times 13''.700 = 7^\circ\ 1'\ 27''.6$$

Since the value of n is nearly one-half, we may interpolate *backwards* from July 9[d] 7[h] with equal facility: thus we find

$$n = 0.567 \qquad \tfrac{n}{2} = 0.283$$

$$\therefore F'_{-\frac{n}{2}} = 13''.582 + 0.283 \times 0''.155 = 13''.626$$

Whence

$$\delta = 7^\circ\ 24'\ 37''.4 - 102 \times 13''.626 = 7^\circ\ 1'\ 27''.55$$

which substantially agrees with the above result.

EXAMPLE II.—From the following table of the moon's horizontal parallax (π), interpolate the value for July $10^{d}\ 16^{h}\ 24^{m}.0$.

Date 1898	Moon's Hor. Parallax	Diff. for 1 Hour	a
d	′ ″	″	″
July 10.0	56 26.1	−2.04	
10.5	56 2.5	1.89	+0.15
11.0	55 40.7	1.73	0.16
11.5	55 21.1	−1.55	+0.18

Here we have

$$T = \text{July } 10^{d}\ 16^{h}.40 \qquad t = \text{July } 10^{d}\ 12^{h}.00$$

$$\omega = 12 \text{ hours} \qquad n = \frac{4^{h}.40}{12^{h}} = 0.367 \qquad \frac{n}{2} = 0.183$$

We therefore obtain

$$F'_{\frac{n}{2}} = -1''.89 + 0.183 \times 0''.16 = -1''.86$$

$$\therefore \pi = 56'\ 2''.5 - 4.4 \times 1''.86 = 55'\ 54''.3$$

Interpolating *backwards* from July $11^{d}\ 0^{h}$, we find

$$\pi = 55'\ 40''.7 + 7.6 \times 1''.78 = 55'\ 54''.2$$

65. *Choice of Formulae in a Given Case.*—When derivatives are required to be computed at or near either the *beginning* or the *end* of a tabular series, the formulae derived from NEWTON's Formula of interpolation must necessarily be employed. In all other cases, the choice lies between STIRLING's and BESSEL's forms, and should be decided by the value of n. When $n = 0$, the formulae (175) are unquestionably the best. When $n = \frac{1}{2}$, the group (187) is especially convenient. As a general rule, subject to change in certain cases, it may be stated that when n lies between the limits 0.25 and 0.75, the formulae derived from BESSEL's Formula of interpolation will be found most convenient: for other values of n, those derived from STIRLING's Formula should be employed.

EXAMPLES.

1. Given the following table of "Latitude Reduction":

φ	$\varphi-\varphi'$	φ	$\varphi-\varphi'$
°	′ ″	°	′ ″
0	0 0.00	15	5 44.32
5	1 59.53	20	7 22.80
10	3 55.47	25	8 47.93

Compute the variation of $\varphi-\varphi'$ corresponding to a change of 10′ in φ, for each of the tabular values of the argument. Denote this variation by v.

2. From the preceding table, find the change in v corresponding to a change of one degree in φ, for $\varphi = 9° 30'$; also for $\varphi = 22° 42'$.

3. The table below contains the obliquity of the ecliptic (ϵ) for every fifth century.

Year, A.D.	ϵ
	° ′ ″
0	23 41 43.78
500	37 57.97
1000	34 8.07
1500	30 15.43
2000	23 26 21.41

Compute the variation of ϵ per century (ϵ') for the years 750 and 1250.

4. From the table of ϵ in Example 3, find the variation of ϵ' per century, for the years 0 and 2000; ϵ' denoting the change in ϵ for 1 century.

5. Given the logarithm of the earth's radius vector ($\log R$) for the following dates:

Date 1898	log R	Date 1898	log R
Dec. 15	9.9930137	Dec. 24	9.9927353
18	9.9929025	27	9.9926858
21	9.9928085	30	9.9926619

Compute the hourly change in log R for the dates Dec. $18^d\ 0^h$, Dec. $22^d\ 12^h$, and Dec. $26^d\ 17^h$. Denote the hourly change by ρ.

6. From the preceding ephemeris of log R, find the daily variation of ρ for the dates Dec. $15^d\ 0^h$, Dec. $24^d\ 0^h$, and Dec. $26^d\ 10^h$.

7. The following table gives the right-ascension of *Mercury*, together with the *hourly difference*, for several alternate days of December, 1898:

Date 1898	R.A. of *Mercury*	Diff. for 1 Hour
	h m s	s
Dec. 1	18 1 2.54	+12.855
3	18 10 50.60	11.587
5	18 19 28.46	9.915
7	18 26 34.57	7.749
9	18 31 43.19	+ 5.009

Compute, by the formulae (200) and (201), the R.A. of *Mercury* for the dates Dec. $4^d\ 14^h\ 22^m.0$ and Dec. $5^d\ 12^h\ 30^m.0$. Check the results by direct interpolation from the tabular right-ascensions.

8. Given the following ephemeris of the moon's right-ascension:

Date 1898	Moon's Right-Ascension	Diff. for 1 Minute
d h	h m s	s
Apr. 8 1	14 27 33.52	2.4508
8 4	14 34 56.35	2.4694
8 7	14 42 22.48	2.4876
8 10	14 49 51.86	2.5054

By the process stated in the rule of § 64, compute the moon's R.A. for the dates Apr. $8^d\ 3^h\ 0^m$; $4^h\ 54^m$; $5^h\ 30^m$; and Apr. $8^d\ 7^h\ 36^m$.

CHAPTER IV.

OF MECHANICAL QUADRATURE.

66. We have shown in the preceding chapter that when a series of equidistant values of any function are known, it is possible to compute special values of the first and higher derivatives of that function, without regard to its analytical form. We shall now consider the inverse problem, namely: *From a series of tabular values of $F(T)$, to find*

$$X = \int_{T'}^{T''} F(T)\,dT$$

where the limits T' and T'' are numerically assigned.

The solution of this important problem is effected by integrating the expression for $F(t+n\omega)$, as given by any one of the several formulae of interpolation, and then giving to n the limiting values which correspond to T' and T''. The method is wholly independent of the analytical form of the function $F(T)$. It is therefore of especial advantage and importance in the following cases:

(*a*) *When the function is analytically unknown.* This is the case with graphical records of continuous observations, so frequently made in physical experiments and tests. As a common example we mention the indicator diagrams of a steam engine. It is usually required to find the area comprised between the "pressure" curve, a fixed base line, and two extreme ordinates. This area may be found, in the generality of cases, by the method proposed.

(*b*) *When the function is analytically known, but is non-integrable.* Under this head are included the most important applications of the method in question. For example, let it be required to find

$$X = \int_{20^\circ}^{52^\circ} \frac{dT}{\sqrt{1-e^2 \sin^2 T}}$$

where e is numerically given. We cannot express the indefinite integral in finite form. If e is sufficiently small (say $e = 0.1$), we may expand $(1-e^2 \sin^2 T)^{-\frac{1}{2}}$ in a series of ascending powers of $e^2 \sin^2 T$, and integrate each term of this expansion separately: a very few terms will then suffice to compute X as accurately as may be required. If, however, the quantity e is nearly equal to unity (say $e = 0.9$), this series does not converge with sufficient rapidity for practical use, and hence the method of expansion fails.

On the other hand, given *any* value of e not exceeding unity, we can readily tabulate $F(T) \equiv (1-e^2 \sin^2 T)^{-\frac{1}{2}}$ for a series of values such as $T = 20°, 24°, 28°, \ldots\ 52°$. Having differenced these values of F, it is then a simple matter to compute X from the numerical data thus furnished. In the nature of the case, however, the process must, in general, be an approximative one; depending, as does the method of interpolation, upon a limited number of (usually approximate) values of the function in question.

The process by which the definite integral of a function is computed from a series of numerical values of that function, is called *mechanical quadrature*, or *numerical integration*. We proceed to develop the formulae which are commonly employed for this purpose.

67. *Quadrature as Based upon* NEWTON'S *Formula of Interpolation.*— Suppose that $i+1$ values of $F(T)$ have been tabulated and differenced as shown in the schedule below:

T	$F(T)$	Δ'	Δ''	Δ'''	Δ^{iv}	Δ^{v}
t	F_0					
		Δ_0'				
$t+\omega$	F_1		Δ_0''			
		Δ_1'		Δ_0'''		
$t+2\omega$	F_2		Δ_1''		Δ_0^{iv}	
		Δ_2'		Δ_1'''		Δ_0^{v}
.	.	.	.	.	.	.
.	.	.	.	.	.	.
.	.	.	.	.	.	.
.	.	.	.	.	.	.
						Δ_{i-5}^{v}
$t+(i-2)\omega$	F_{i-2}		Δ_{i-3}''	Δ_{i-3}'''	Δ_{i-4}^{iv}	
		Δ'_{i-2}				
$t+(i-1)\omega$	F_{i-1}		Δ_{i-2}''			
		Δ'_{i-1}				
$t+i\omega$	F_i					

Let it be required to find from this table the value of

$$X = \int_t^{t+i\omega} F(T)\, dT \tag{209}$$

Since

we have
$$\left. \begin{aligned} T &= t + n\omega \\ dT &= \omega dn \end{aligned} \right\} \tag{210}$$

and therefore
$$X = \int_t^{t+\omega} F(T)\, dT = \omega \int_0^1 F(t+n\omega)\, dn \tag{211}$$

Now, by NEWTON'S Formula, we have
$$F(t+n\omega) = F_0 + n\Delta_0' + B\Delta_0'' + C\Delta_0''' + D\Delta_0^{\text{iv}} + \ldots$$

where B, C, D, ... denote the binomial coefficients of the nth order. Multiplying by dn, and integrating, we obtain
$$\int F(t+n\omega)\, dn = \int (F_0 + n\Delta_0' + B\Delta_0'' + C\Delta_0''' + \ldots)\, dn$$

or
$$\int F(t+n\omega)\, dn = nF_0 + \tfrac{n^2}{2}\Delta_0' + \Delta_0'' \int B dn + \Delta_0''' \int C dn + \ldots + M \tag{212}$$

where M is the constant of integration. If, for brevity, we put
$$\beta = \int_0^1 B dn \quad , \quad \gamma = \int_0^1 C dn \quad , \quad \delta = \int_0^1 D dn \; , \tag{213}$$

then, from the preceding equation, we derive
$$\int_0^1 F(t+n\omega)\, dn = F_0 + \tfrac{1}{2}\Delta_0' + \beta\Delta_0'' + \gamma\Delta_0''' + \delta\Delta_0^{\text{iv}} + \ldots \tag{214}$$

Whence we obtain, in succession,
$$\left. \begin{aligned} \int_1^2 F(t+n\omega)\, dn &= \int_0^1 F(t+\omega+n\omega)\, dn = F_1 + \tfrac{1}{2}\Delta_1' + \beta\Delta_1'' + \gamma\Delta_1''' + \ldots \\ \int_2^3 F(t+n\omega)\, dn &= \int_0^1 F(t+2\omega+n\omega) dn = F_2 + \tfrac{1}{2}\Delta_2' + \beta\Delta_2'' + \gamma\Delta_2''' + \ldots \\ &\ldots\ldots\ldots\ldots\ldots\ldots \\ \int_{i-1}^i F(t+n\omega)\, dn &= \int_0^1 F(t+\overline{i-1}\omega+n\omega)\, dn = F_{i-1} + \tfrac{1}{2}\Delta_{i-1}' + \beta\Delta_{i-1}'' + \gamma\Delta_{i-1}''' + \ldots \end{aligned} \right\} \tag{215}$$

Summing the integrals expressed in (214) and (215), we find
$$\int_0^i F(t+n\omega)\, dn = \sum_{r=0}^{r=i-1} F_r + \tfrac{1}{2}\sum_{r=0}^{r=i-1}\Delta_r' + \beta\sum_{r=0}^{r=i-1}\Delta_r'' + \gamma\sum_{r=0}^{r=i-1}\Delta_r''' + \ldots \tag{216}$$

The numerical values of β, γ, δ, ... (sometimes called the *coefficients of quadrature*) must now be determined. These may be found directly by integrating the expressions for B, C, D, ... , as expanded in (163), and then taking the limits of n according to (213). But the following indirect method seems preferable, since it adds a significance to the result. Let us put

$$Q = \int (1+y)^n \, dn = \int (1+ny+By^2+Cy^3+Dy^4+ \; . \; . \; . \; . \;) \, dn \tag{217}$$

where y is supposed constant. Then, if we also put

$$Q' = \int_0^1 (1+y)^n \, dn$$

we shall have

$$Q' = 1 + \tfrac{1}{2} y + \beta y^2 + \gamma y^3 + \delta y^4 + \epsilon y^5 + \zeta y^6 + \; . \; . \; . \; . \tag{218}$$

the coefficients being those defined in (213).

Again, put

$$(1+y)^n = z \tag{219}$$

that is

$$n \log (1+y) = \log z$$

and we find

$$\log (1+y) \; . \; dn = \frac{dz}{z}$$

or

$$z dn = \frac{dz}{\log (1+y)} \tag{220}$$

We therefore obtain

$$Q = \int (1+y)^n \, dn = \int z dn = \int \frac{dz}{\log (1+y)} = \frac{z}{\log (1+y)} + \text{const.} = \frac{(1+y)^n}{\log (1+y)} + \text{const.}$$

Whence

$$Q' = \int_0^1 (1+y)^n dn = \left[\frac{(1+y)^n}{\log (1+y)} \right]_{n=0}^{n=1} = \frac{y}{\log (1+y)}$$

$$= \frac{y}{y - \frac{y^2}{2} + \frac{y^3}{3} - \frac{y^4}{4} + \; . \; . \; . \; .} = \left(1 - \frac{y}{2} + \frac{y^2}{3} - \frac{y^3}{4} + \frac{y^4}{5} - \; . \; . \; . \; . \; \right)^{-1}$$

Expanding the last expression by the Binomial Theorem, or by direct division, we obtain

$$Q' = 1 + \tfrac{1}{2} y - \tfrac{1}{12} y^2 + \tfrac{1}{24} y^3 - \tfrac{19}{720} y^4 + \tfrac{3}{160} y^5 - \tfrac{863}{60480} y^6 + \; . \; . \; . \; . \tag{221}$$

Whence, comparing (218) and (221), we find

$$\left. \begin{array}{ll} \beta = -\tfrac{1}{12} & \epsilon = +\tfrac{3}{160} \\ \gamma = +\tfrac{1}{24} & \zeta = -\tfrac{863}{60480} \\ \delta = -\tfrac{19}{720} & \; . \; . \; . \; . \; . \; . \end{array} \right\} \tag{222}$$

which are the numerical values of the coefficients of formula (216). It therefore appears that the fundamental *coefficients of quadrature* are those in the expansion of $[\log(1+y)]^{-1}$.

Let us now regard the functions $F_0, F_1, F_2, \ldots F_i$ as first differences of an auxiliary functional series which we shall designate $'F$. A schedule containing the new series may be conveniently arranged as follows:

T	$'F$	$F(T)$	Δ'	Δ''	Δ'''
	$'F_0$				
t		F_0			
	$'F_1$		Δ_0'		
$t+\omega$		F_1		Δ_0''	
	$'F_2$		Δ_1'		Δ_0'''
$t+2\omega$		F_2		Δ_1''	
	$'F_3$		.		. .
.		.		.	
	.		.		.
.		.		.	
	.		.		.
.		.		.	
	$'F_{i-1}$		Δ'_{i-2}		Δ'''_{i-3}
$t+(i-1)\omega$		F_{i-1}		Δ''_{i-2}	
	$'F_i$		Δ'_{i-1}		
$t+i\omega$		F_i			
	$'F_{i+1}$				

The value of $'F_0$ is entirely arbitrary. Having assigned a convenient value to this quantity, the remaining terms in the series are readily formed by successive additions, thus:

$$'F_1 = {}'F_0 + F_0 \;, \quad 'F_2 = {}'F_1 + F_1 \;, \quad \ldots\ldots \;, \quad 'F_{i+1} = {}'F_i + F_i$$

We shall now put the second member of (216) under a form more convenient for computation. By Theorem I, we have

$$\left.\begin{aligned}
\sum_{r=0}^{r=i-1} F_r &\equiv F_0 + F_1 + F_2 + \ldots + F_{i-1} = {}'F_i - {}'F_0 \\
\sum_{r=0}^{r=i-1} \Delta_r' &\equiv \Delta_0' + \Delta_1' + \Delta_2' + \ldots + \Delta'_{i-1} = F_i - F_0 \\
\sum_{r=0}^{r=i-1} \Delta_r'' &\equiv \Delta_0'' + \Delta_1'' + \Delta_2'' + \ldots + \Delta''_{i-1} = \Delta_i' - \Delta_0' \\
\sum_{r=0}^{r=i-1} \Delta_r''' &\equiv \Delta_0''' + \Delta_1''' + \Delta_2''' + \ldots + \Delta'''_{i-1} = \Delta_i'' - \Delta_0'' \\
&\ldots\ldots\ldots\ldots\ldots\ldots
\end{aligned}\right\} \qquad (223)$$

and hence (216) becomes

$$\int_0^i F(t+n\omega)\,dn = ({}'F_i - {}'F_0) + \tfrac{1}{2}(F_i - F_0) + \beta(\Delta_i' - \Delta_0') + \gamma(\Delta_i'' - \Delta_0'') + \delta(\Delta_i''' - \Delta_0''') + \epsilon(\Delta_i^{iv} - \Delta_0^{iv}) + \ldots \qquad (224)$$

This formula possesses the disadvantage of involving differences Δ_i', Δ_i'', Δ_i''', ... which are not furnished by the foregoing schedule. To obviate this difficulty, we proceed as follows:

Put

$$q = {}'F_i + \tfrac{1}{2}F_i + \beta\Delta_i' + \gamma\Delta_i'' + \delta\Delta_i''' + \epsilon\Delta_i^{\text{iv}} + \zeta\Delta_i^{\text{v}} + \dots \tag{225}$$

and (224) may then be written

$$\int_0^1 F(t+n\omega)\,dn = q - ({}'F_0 + \tfrac{1}{2}F_0 + \beta\Delta_0' + \gamma\Delta_0'' + \delta\Delta_0''' + \dots) \tag{226}$$

Upon giving to n, in formula (75), the values $+1$, 0, -1, -2, -3, -4, ..., successively, we obtain

$$\left.\begin{aligned}
{}'F_i &= {}'F_{i+1} - F_i \\
F_i &= F_i \\
\Delta_i' &= \Delta'_{i-1} + \Delta''_{i-2} + \Delta'''_{i-3} + \Delta^{\text{iv}}_{i-4} + \Delta^{\text{v}}_{i-6} + \dots \\
\Delta_i'' &= \Delta''_{i-2} + 2\Delta'''_{i-3} + 3\Delta^{\text{iv}}_{i-4} + 4\Delta^{\text{v}}_{i-5} + \dots \\
\Delta_i''' &= \Delta'''_{i-3} + 3\Delta^{\text{iv}}_{i-4} + 6\Delta^{\text{v}}_{i-5} + \dots \\
\Delta_i^{\text{iv}} &= \Delta^{\text{iv}}_{i-4} + 4\Delta^{\text{v}}_{i-5} + \dots \\
&\dots\dots\dots\dots\dots\dots
\end{aligned}\right\} \tag{227}$$

If these expressions be substituted in (225), we shall have q in terms of the known tabular differences, and hence obtain the required integral from (226). To avoid the labor of numerical reduction incident to this substitution, we derive the result in the following indirect manner: Put

$$\theta = \frac{1}{\log(1+x)} = x^{-1} + \tfrac{1}{2}x^0 + \beta x + \gamma x^2 + \delta x^3 + \epsilon x^4 + \zeta x^5 + \dots \tag{228}$$

Also, take

$$x = \frac{u}{1-u} \tag{229}$$

and we have

$$\left.\begin{aligned}
x^{-1} &= u^{-1}(1-u) = u^{-1} - u^0 \\
x^0 &= u^0 \\
x &= u\,(1-u)^{-1} = u + u^2 + u^3 + u^4 + u^5 + \dots \\
x^2 &= u^2(1-u)^{-2} = u^2 + 2u^3 + 3u^4 + 4u^5 + \dots \\
x^3 &= u^3(1-u)^{-3} = u^3 + 3u^4 + 6u^5 + \dots \\
x^4 &= u^4(1-u)^{-4} = u^4 + 4u^5 + \dots \\
&\dots\dots\dots\dots\dots\dots
\end{aligned}\right\} \tag{230}$$

If now we substitute these expressions for x^{-1}, x^0, x, x^2, ... in the second member of (228), we obtain θ in terms of u^{-1}, u^0, u, u^2, ... But it will be observed that this operation is identical in algebraic form with the substitution above proposed with respect to (227) and (225); for the θ operation involves the quantities

$$\theta;\quad x^{-1},\ x^0,\ x,\ x^2,\ x^3,\ \ldots\ ;\quad u^{-1},\ u^0,\ u,\ u^2,\ u^3,\ \ldots\ ;$$

while the q operation involves, in precisely the same algebraic relations, the quantities

$$q;\quad {}'F_i,\ F_i,\ \varDelta_i',\ \varDelta_i'',\ \varDelta_i''',\ \ldots\ ;\quad {}'F_{i+1},\ F_i,\ \varDelta'_{i-1},\ \varDelta''_{i-2},\ \varDelta'''_{i-3},\ \ldots$$

Hence the result for q will immediately follow when the result for θ has been derived. But we may obtain θ as a function of u, in the form required, more simply than by direct substitution of the expressions (230) in (228). For, by (229), we have

$$1+x = \frac{1}{1-u}$$

whence

$$\log(1+x) = -\log(1-u) \tag{231}$$

Therefore, by (228), we find

$$\theta = \frac{1}{\log(1+x)} = -\frac{1}{\log(1-u)} = u^{-1} - \tfrac{1}{2}u^0 + \beta u - \gamma u^2 + \delta u^3 - \epsilon u^4 + \zeta u^5 - \ldots \tag{232}$$

Accordingly, writing q for θ, ${}'F_{i+1}$ for u^{-1}, F_i for u^0, $\varDelta'_{i-1}$ for u, etc., as justified by the preceding reasoning, we obtain

$$q = {}'F_{i+1} - \tfrac{1}{2}F_i + \beta\varDelta'_{i-1} - \gamma\varDelta''_{i-2} + \delta\varDelta'''_{i-3} - \epsilon\varDelta^{\text{iv}}_{i-4} + \zeta\varDelta^{\text{v}}_{i-5} - \ldots \tag{233}$$

Substituting this value of q in (226), and grouping like terms, we get

$$\begin{aligned}\int_0^i F(t+n\omega)\,dn = {} & ({}'F_{i+1} - {}'F_0) - \tfrac{1}{2}(F_i + F_0) + \beta(\varDelta'_{i-1} - \varDelta'_0) \\ & - \gamma(\varDelta''_{i-2} + \varDelta_0'') + \delta(\varDelta'''_{i-3} - \varDelta_0''') - \epsilon(\varDelta^{\text{iv}}_{i-4} + \varDelta_0^{\text{iv}}) + \ldots\end{aligned} \tag{234}$$

Whence, restoring the values of β, γ, δ, ... , as given in (222), and applying (211), we have

$$\begin{aligned}\int_t^{t+i\omega} F(T)\,dT &= \omega\int_0^i F(t+n\omega)\,dn \\ &= \omega\{({}'F_{i+1} - {}'F_0) - \tfrac{1}{2}(F_i + F_0) - \tfrac{1}{12}(\varDelta'_{i-1} - \varDelta_0') - \tfrac{1}{24}(\varDelta''_{i-2} + \varDelta_0'') \\ &\quad - \tfrac{19}{720}(\varDelta'''_{i-3} - \varDelta_0''') - \tfrac{3}{160}(\varDelta^{\text{iv}}_{i-4} + \varDelta_0^{\text{iv}}) - \tfrac{863}{60480}(\varDelta^{\text{v}}_{i-5} - \varDelta_0^{\text{v}}) - \ldots\}\end{aligned} \tag{235}$$

When the tabulation of the function extends beyond the value F_i, it is sometimes more convenient to employ the following formula, easily obtained from (224):

$$\int_t^{t+i\omega} F(T)\,dT = \omega\int_0^i F(t+n\omega)\,dn$$
$$= \omega\{('F_i-'F_0)+\tfrac{1}{2}(F_i-F_0)-\tfrac{1}{12}(\Delta_i'-\Delta_0')+\tfrac{1}{24}(\Delta_i''-\Delta_0'')$$
$$-\tfrac{19}{720}(\Delta_i'''-\Delta_0''')+\tfrac{3}{160}(\Delta_i^{\text{iv}}-\Delta_0^{\text{iv}})-\tfrac{863}{60480}(\Delta_i^{\text{v}}-\Delta_0^{\text{v}})+\;.\;.\;.\;.\} \qquad (236)$$

We here emphasize the fact that the value of $'F_0$ is wholly arbitrary.

68. As an example in the use of formula (235), let it be required to find*

$$X = \int_{20°}^{44°} \cos T\,dT$$

using six places of decimals.

The first consideration concerns the tabular interval to be employed. It is desirable to tabulate as few values of the function as are consistent with a convenient schedule of differences. In all cases the differences should sensibly vanish beyond the third or fourth order. Adopting $\omega = 4°$ as a suitable interval in the present instance, we obtain the following table of $F(T) \equiv \cos T$:

T	$'F$	$F(T)\equiv\cos T$	Δ'	Δ''	Δ'''	Δ^{iv}
	0.000000					
20°		0.939693				
	0.939693		−26148			
24		0.913545		−4449		
	1.853238		30597		+146	
28		0.882948		4303		+26
	2.736186		34900		172	
32		0.848048		4131		17
	3.584234		39031		189	
36		0.809017		3942		+22
	4.393251		42973		+211	
40		0.766044		−3731		
	5.159295		−46704			
44		0.719340				
	5.878635					

Taking $t = 20°$, and assuming the arbitrary quantity $'F_0 = 0$, we complete the column $'F$ by successive additions. Whence, by (235), we find

* In selecting examples of numerical integration for the present chapter, we have in most cases chosen for $F(T)$ some simple, *integrable* function, whose tabular values are readily taken or formed from various tables in common use. By such selection we gain in simplicity, while losing little or nothing of generality; and, moreover, from thus knowing *a priori* the true value of the integral sought, we are at once informed as to the final accuracy of each application.

$(i = 6)$		$'F_7 - 'F_0$	$= +5.878635$
$F_6 + F_0$	$= +1.659033$	$- \frac{1}{2}(F_6 + F_0)$	$= -0.829516.5$
$\varDelta_5' - \varDelta_0'$	$= -\ \ 20556$	$- \frac{1}{12}(\varDelta_5' - \varDelta_0')$	$= +\ \ 1713.0$
$\varDelta_4'' + \varDelta_0''$	$= -\ \ 8180$	$- \frac{1}{24}(\varDelta_4'' + \varDelta_0'')$	$= +\ \ 340.8$
$\varDelta_3''' - \varDelta_0'''$	$= +\ \ 65$	$- \frac{19}{720}(\varDelta_3''' - \varDelta_0''')$	$= -\ \ 1.7$
$\varDelta_2^{iv} + \varDelta_0^{iv}$	$= +\ \ 48$	$- \frac{3}{160}(\varDelta_2^{iv} + \varDelta_0^{iv})$	$= -\ \ 0.9$
$\log \Sigma =$	0.703392	sum, Σ	$= +5.051170$
$\log \omega =$	8.843937	ω	$= 4° = \frac{\pi}{45}$
$\log X =$	9.547329	$\therefore X$	$= 0.352638$

Since $\int \cos T dT = \sin T$, we find for the true value of the definite integral,

$$X = \sin 44° - \sin 20°$$
$$= 0.694658 - 0.342020 = 0.352638$$

If it be required to compute

$$X = \int_{20°}^{28°} \cos T dT$$

from the foregoing table, formula (236) at once serves the purpose. Thus we obtain

$(i = 2)$		$'F_2 - 'F_0$	$= +1.853238$
$F_2 - F_0$	$= -56745$	$+ \frac{1}{2}(F_2 - F_0)$	$= -\ \ 28372.5$
$\varDelta_2' - \varDelta_0'$	$= -\ \ 8752$	$- \frac{1}{12}(\varDelta_2' - \varDelta_0')$	$= +\ \ 729.3$
$\varDelta_2'' - \varDelta_0''$	$= +\ \ 318$	$+ \frac{1}{24}(\varDelta_2'' - \varDelta_0'')$	$= +\ \ 13.3$
$\varDelta_2''' - \varDelta_0'''$	$= +\ \ 43$	$- \frac{19}{720}(\varDelta_2''' - \varDelta_0''')$	$= -\ \ 1.1$
		Σ	$= +1.825607$
		$\therefore X$	$= 0.127451$

Here the true value evidently is —

$$X = \sin 28° - \sin 20° = 0.127451$$

69. *Precepts for Computing the Definite Integral when One or Both Limits Fail to Coincide with some Tabular Value of the Argument T.* — Thus far we have considered the limits of the integral

$$X = \int_{T'}^{T''} F(T)\, dT$$

to be of the form

$$T' = t + i'\omega \quad , \quad T'' = t + i''\omega$$

where i' and i'' are integers, and hence T' and T'' are two particular values of T for which $F(T)$ has been tabulated. We shall now consider the more general problem of finding X when the limits have the form

$$T' = t + n'\omega \quad , \quad T'' = t + n''\omega$$

where n' and n'' are non-integers — that is, either proper fractions or mixed numbers.

To illustrate the significance of the problem in question, suppose it were required to find by mechanical quadrature the value of

$$X = \int_{21°\,13'\,37''}^{42°\,46'\,54''} \cos T dT$$

Obviously, it would be impracticable to tabulate the function for a series of equidistant values of T, of which $T' = 21°\ 13'\ 37''$ and $T'' = 42°\ 46'\ 54''$ are two particular terms. We may, however, employ the same table as was used in the preceding examples, constructed for $T = 20°, 24°, 28°, \ldots\ 44°$, and obtain the required result by *interpolation*. Thus, in the examples just mentioned, we have computed the values of X from the lower limit $T' = 20°$ to the upper limits $T'' = 44°$ and $28°$, respectively. In like manner, keeping the lower limit always $= 20°$, we may find the integral corresponding to each of the following values of the upper limit, viz.:

$$T'' = 20°,\ 24°,\ 28°,\ \ldots\ 44°, \text{ respectively};$$

that is, for each of the tabular values of T. Then, having differenced the resulting values of the integral, we may readily find *by interpolation* the values which correspond to the upper limits $21°\ 13'\ 37''$ and $42°\ 46'\ 54''$. Denoting these interpolated values by X' and X'' respectively, we have

$$X' = \int_{20°}^{21°\,13'\,37''} \cos T dT \quad , \quad X'' = \int_{20°}^{42°\,46'\,54''} \cos T dT$$

and therefore

$$X = \int_{21°\,13'\,37''}^{42°\,46'\,54''} \cos T dT = X'' - X'$$

We leave the detailed solution of this example to the student as a valuable exercise, exhibiting the spirit of the method employed in problems of this type. The process actually used differs somewhat in form from the method here explained; but the principle remains the same. We proceed to develop the general formulae.

70. Let us put

$$I_i = \int_0^i F(t+n\omega)\,dn \tag{237}$$

and

$$\Psi(i) = {}'F_i + \tfrac{1}{2}F_i + \beta\Delta_i' + \gamma\Delta_i'' + \delta\Delta_i''' + \epsilon\Delta_i^{\text{iv}} + \ldots \tag{238}$$

where i denotes an integer. Then (224) becomes

$$I_i = \Psi(i) - \Psi(0) \tag{239}$$

Let us now suppose that (239) has been computed for $i = 0, 1, 2, 3, 4, \ldots$, in succession. Then, from the series of values

$$\left.\begin{aligned} I_0 &= \Psi(0) - \Psi(0) \\ I_1 &= \Psi(1) - \Psi(0) \\ I_2 &= \Psi(2) - \Psi(0) \\ &\ldots\ldots\ldots \end{aligned}\right\} \tag{240}$$

thus determined, it is evident that any *intermediate* value, say I_n, can be found by interpolation. To derive a general formula for this purpose, we must express the differences of the series (240). Now, by (238), we have

$$\left.\begin{aligned} \Psi(0) &= {}'F_0 + \tfrac{1}{2}F_0 + \beta\Delta_0' + \gamma\Delta_0'' + \delta\Delta_0''' + \epsilon\Delta_0^{\text{iv}} + \ldots \\ \Psi(1) &= {}'F_1 + \tfrac{1}{2}F_1 + \beta\Delta_1' + \gamma\Delta_1'' + \delta\Delta_1''' + \epsilon\Delta_1^{\text{iv}} + \ldots \\ \Psi(2) &= {}'F_2 + \tfrac{1}{2}F_2 + \beta\Delta_2' + \gamma\Delta_2'' + \delta\Delta_2''' + \epsilon\Delta_2^{\text{iv}} + \ldots \\ &\ldots\ldots\ldots\ldots\ldots\ldots\ldots\ldots \end{aligned}\right\} \tag{241}$$

whence, observing the general relation

$$\Delta_{i+1}^{(r)} - \Delta_i^{(r)} = \Delta_i^{(r+1)}$$

we derive the following schedule of differences:

Function	1st Differences	2d Differences	3d Differences
$I_0 = \Psi(0) - \Psi(0)$	$F_0 + \frac{1}{2}\Delta_0' + \beta\Delta_0'' + \gamma\Delta_0''' + \ldots$	$\Delta_0' + \frac{1}{2}\Delta_0'' + \beta\Delta_0''' + \ldots$	$\Delta_0'' + \frac{1}{2}\Delta_0''' + \ldots$
$I_1 = \Psi(1) - \Psi(0)$	$F_1 + \frac{1}{2}\Delta_1' + \beta\Delta_1'' + \gamma\Delta_1''' + \ldots$	$\Delta_1' + \frac{1}{2}\Delta_1'' + \beta\Delta_1''' + \ldots$	$\Delta_1'' + \frac{1}{2}\Delta_1''' + \ldots$
$I_2 = \Psi(2) - \Psi(0)$	$F_2 + \frac{1}{2}\Delta_2' + \beta\Delta_2'' + \gamma\Delta_2''' + \ldots$	$\Delta_2' + \frac{1}{2}\Delta_2'' + \beta\Delta_2''' + \ldots$	…………
$I_3 = \Psi(3) - \Psi(0)$	…………………	…………………	
…………			

Therefore, applying NEWTON's Formula of interpolation, we have

$$\begin{aligned} I_n &= I_0 + n\,(\text{1st Diff.}) + B\,(\text{2d Diff.}) + C\,(\text{3d Diff.}) + \ldots \\ &= \Psi(0) - \Psi(0) + n\,(F_0 + \tfrac{1}{2}\Delta_0' + \beta\Delta_0'' + \gamma\Delta_0''' + \ldots) \\ &\quad + B\,(\Delta_0' + \tfrac{1}{2}\Delta_0'' + \beta\Delta_0''' + \ldots) + C\,(\Delta_0'' + \tfrac{1}{2}\Delta_0''' + \ldots) + D\,(\Delta_0''' + \ldots) + \ldots \end{aligned}$$

By transposing the term $-\Psi(0)$ to the first member, and substituting for $\Psi(0)$ in the second member the expression given by (241), we find

$$\begin{aligned} I_n + \Psi(0) = &\ ('F_0 + \tfrac{1}{2} F_0 + \beta \Delta_0' + \gamma \Delta_0'' + \delta \Delta_0''' + \dots) \\ &+ n(F_0 + \tfrac{1}{2}\Delta_0' + \beta \Delta_0'' + \gamma \Delta_0''' + \dots) \\ &+ B(\Delta_0' + \tfrac{1}{2}\Delta_0'' + \beta \Delta_0''' + \dots) + C(\Delta_0'' + \tfrac{1}{2}\Delta_0''' + \dots) + D(\Delta_0''' + \dots) + \dots \end{aligned}$$

Upon arranging the last expression according to the coefficients $1, \frac{1}{2}, \beta, \gamma, \delta, \dots,$ it becomes

$$\begin{aligned} I_n + \Psi(0) = &\ ('F_0 + nF_0 + B\Delta_0' + C\Delta_0'' + D\Delta_0''' + \dots) \\ &+ \tfrac{1}{2}(F_0 + n\Delta_0' + B\Delta_0'' + C\Delta_0''' + \dots) \\ &+ \beta(\Delta_0' + n\Delta_0'' + B\Delta_0''' + \dots) \\ &+ \gamma(\Delta_0'' + n\Delta_0''' + \dots) \\ &+ \delta(\Delta_0''' + \dots) \\ &+ \dots \end{aligned}$$

Now, it will be observed that the first polynomial in the second member of this equation is simply the expression for $'F_n$, — the quantity derived from the series $'F_0, 'F_1, 'F_2, \dots$ by interpolation. Similarly, the remaining parentheses contain the expressions for $F_n, \Delta'_n, \Delta_n''$, $\dots,$ likewise derived by interpolation from their respective series. We therefore have

$$I_n + \Psi(0) = 'F_n + \tfrac{1}{2} F_n + \beta\Delta'_n + \gamma\Delta_n'' + \delta\Delta_n''' + \dots = \Psi(n) \tag{242}$$

Whence

$$\int_0^n F(t+n\omega)\,dn = I_n = \Psi(n) - \Psi(0) \tag{243}$$

71. In like manner, if we put

$$\varphi(i) = 'F_{i+1} - \tfrac{1}{2}F_i + \beta\Delta'_{i-1} - \gamma\Delta''_{i-2} + \delta\Delta'''_{i-3} - \dots \tag{244}$$

then, by (234), we have

$$I_i = \int_0^i F(t+n\omega)\,dn = \varphi(i) - \Psi(0)$$

Therefore, by interpolation (reasoning precisely as above), we obtain

$$\int_0^n F(t+n\omega)\,dn = \varphi(n) - \Psi(0) \tag{245}$$

Again, writing n' for the upper limit n in (243), and n'' for n in (245), we get

$$\int_0^{n'} F(t+n\omega)\,dn = \Psi(n') - \Psi(0) \quad , \quad \int_0^{n''} F(t+n\omega)\,dn = \varphi(n'') - \Psi(0)$$

the difference of which gives

$$\int_{n'}^{n''} F(t+n\omega)\,dn = \varphi(n'') - \Psi(n') \tag{246}$$

Upon substituting in equations (243) and (245) the expressions for Ψ and φ as given by (238) and (244), and restoring the numerical values of β, γ, δ, ... from (222), we obtain

$$\begin{aligned}\int_t^{t+n\omega} F(T)\,dT &= \omega\int_0^n F(t+n\omega)\,dn \\ &= \omega\{('F_n - 'F_0) + \tfrac{1}{2}(F_n - F_0) - \tfrac{1}{12}(\Delta'_n - \Delta'_0) + \tfrac{1}{24}(\Delta_n'' - \Delta_0'') \\ &\quad - \tfrac{19}{720}(\Delta_n''' - \Delta_0''') + \tfrac{3}{160}(\Delta_n^{\mathrm{iv}} - \Delta_0^{\mathrm{iv}}) - \tfrac{863}{60480}(\Delta_n^{\mathrm{v}} - \Delta_0^{\mathrm{v}}) + \; . \; . \; . \; . \}\end{aligned} \tag{247}$$

$$\begin{aligned}\int_t^{t+n\omega} F(T)\,dT &= \omega\int_0^n F(t+n\omega)\,dn \\ &= \omega\{('F_{n+1} - 'F_0) - \tfrac{1}{2}(F_n + F_0) - \tfrac{1}{12}(\Delta'_{n-1} - \Delta'_0) - \tfrac{1}{24}(\Delta''_{n-2} + \Delta_0'') \\ &\quad - \tfrac{19}{720}(\Delta'''_{n-3} - \Delta_0''') - \tfrac{3}{160}(\Delta^{\mathrm{iv}}_{n-4} + \Delta_0^{\mathrm{iv}}) - \tfrac{863}{60480}(\Delta^{\mathrm{v}}_{n-5} - \Delta_0^{\mathrm{v}}) - \; . \; . \; . \; . \}\end{aligned} \tag{248}$$

In like manner, we derive from (246),

$$\begin{aligned}\int_{t+n'\omega}^{t+n''\omega} F(T)\,dT &= \omega\int_{n'}^{n''} F(t+n\omega)\,dn \\ &= \omega\{('F_{n''+1} - 'F_{n'}) - \tfrac{1}{2}(F_{n''} + F_{n'}) - \tfrac{1}{12}(\Delta'_{n''-1} - \Delta'_{n'}) - \tfrac{1}{24}(\Delta''_{n''-2} + \Delta''_{n'}) \\ &\quad - \tfrac{19}{720}(\Delta'''_{n''-3} - \Delta'''_{n'}) - \tfrac{3}{160}(\Delta^{\mathrm{iv}}_{n''-4} + \Delta^{\mathrm{iv}}_{n'}) - \tfrac{863}{60480}(\Delta^{\mathrm{v}}_{n''-5} - \Delta^{\mathrm{v}}_{n'}) - \; . \; . \; . \}\end{aligned} \tag{249}$$

In these formulae the quantities n, n' and n'' are either proper fractions or mixed numbers; while the value of $'F_0$ is wholly arbitrary.

It frequently happens that we have to compute

$$X = \int_t^T F(T)\,dT$$

for several different values of T; the lower limit remaining fixed and equal to t. In such cases it is convenient to determine the arbitrary quantity $'F_0$, in (247) and (248), such that the sum of the terms having the subscript *zero* will vanish. Accordingly, we may arrange these formulae as follows :

Take

$$'F_0 = -\tfrac{1}{2}F_0 + \tfrac{1}{12}\Delta_0' - \tfrac{1}{24}\Delta_0'' + \tfrac{19}{720}\Delta_0''' - \tfrac{3}{160}\Delta_0^{\text{iv}} + \tfrac{863}{60480}\Delta_0^{\text{v}} - \ldots\ldots$$

Then —

(*a*) *When the upper limit falls near the beginning or middle of the tabular series*, find

$$\int_t^{t+n\omega} F(T)\,dT = \omega\int_0^n F(t+n\omega)\,dn$$

$$= \omega('F_n + \tfrac{1}{2}F_n - \tfrac{1}{12}\Delta'_n + \tfrac{1}{24}\Delta_n'' - \tfrac{19}{720}\Delta_n''' + \tfrac{3}{160}\Delta_n^{\text{iv}} - \tfrac{863}{60480}\Delta_n^{\text{v}} + \ldots\ldots) \qquad (250)$$

(*b*) *When the upper limit falls near the end of the series*, find

$$\int_t^{t+n\omega} F(T)\,dT = \omega\int_0^n F(t+n\omega)\,dn$$

$$= \omega('F_{n+1} - \tfrac{1}{2}F_n - \tfrac{1}{12}\Delta'_{n-1} - \tfrac{1}{24}\Delta''_{n-2} - \tfrac{19}{720}\Delta'''_{n-3} - \tfrac{3}{160}\Delta^{\text{iv}}_{n-4} - \tfrac{863}{60480}\Delta^{\text{v}}_{n-5} - \ldots)$$

EXAMPLE I.—Let it be required to find

$$X = \int_{0.42737}^{0.53064} \frac{10\,dT}{\sqrt{T(1-T)}}$$

Here we adopt the interval $\omega = 0.02$, and proceed to form a table for $T = 0.42$, 0.44, 0.46, ... 0.54. Instead of tabulating the *given* function, it is more expedient to tabulate ω times this quantity. All differences are thus multiplied by the same factor, and hence the final multiplication by ω is avoided. We therefore compute

$$F(T) \equiv 0.02 \times \frac{10}{\sqrt{T(1-T)}} = \frac{0.2}{\sqrt{T(1-T)}}$$

for the values of T given above. The result is as follows:

T	$'F$	$F(T) \equiv \frac{0.2}{\sqrt{T(1-T)}}$	Δ'	Δ''	Δ'''	Δ^{iv}
	0.000000					
0.42		0.405220				
	0.405220		−2308			
0.44		0.402912		+682		
	0.808132		1626		−22	
0.46		0.401286		660		+ 8
	1.209418		966		14	
0.48		0.400320		646		8
	1.609738		− 320		− 6	
0.50		0.400000		640		+12
	2.009738		+ 320		+ 6	
0.52		0.400320		+646		
	2.410058		+ 966			
0.54		0.401286				
	2.811344					

The computation is now readily effected by formula (249). Taking $t = 0.42$, we make $'F_0 = 0$, and complete the auxiliary series $'F$. For the values of n' and n'', we have

$$n' = \frac{0.42737 - 0.42}{0.02} = 0.3685$$

$$n'' = \frac{0.53054 - 0.42}{0.02} = 5.5270 = 6 - 0.4730$$

Whence, interpolating by Newton's Formula, we obtain

$$\begin{aligned} 'F_{n'} &= +0.149636.4 & \qquad 'F_{n''+1} &= +2.621373.8 \\ F_{n'} &= +0.404288 & F_{n''} &= +0.400748 \\ \Delta'_{n'} &= - \quad 2054 & \Delta'_{n''-1} &= + \quad 659 \\ \Delta''_{n'} &= + \quad 673 & \Delta''_{n''-2} &= + \quad 642 \\ \Delta'''_{n'} &= - \quad 19 & \Delta'''_{n''-3} &= \quad 0 \end{aligned}$$

Accordingly, by (249), we find

$$\begin{aligned} & & \qquad ('F_{n''+1} - 'F_{n'}) &= +2.471737.4 \\ F_{n''} + F_{n'} &= +0.805036 & -\tfrac{1}{2}\,(F_{n''} + F_{n'}) &= -0.402518.0 \\ \Delta'_{n''-1} - \Delta'_{n'} &= + \quad 2713 & -\tfrac{1}{12}\,(\Delta'_{n''-1} - \Delta'_{n'}) &= - \quad 226.1 \\ \Delta''_{n''-2} + \Delta''_{n'} &= + \quad 1315 & -\tfrac{1}{24}\,(\Delta''_{n''-2} + \Delta''_{n'}) &= - \quad 54.8 \\ \Delta'''_{n''-3} - \Delta'''_{n'} &= + \quad 19 & -\tfrac{19}{720}\,(\Delta'''_{n''-3} - \Delta'''_{n'}) &= - \quad 0.5 \\ \hline & & \therefore X &= +2.068938 \end{aligned}$$

To verify this result, we observe that

$$\int \frac{dT}{\sqrt{T(1-T)}} = 2 \sin^{-1} \sqrt{T}$$

and therefore

$$\begin{aligned} X &= 20\,(\sin^{-1}\sqrt{0.53054} - \sin^{-1}\sqrt{0.42737}) \\ &= 20\,(168303''.25 - 146965''.80) \sin 1'' = 2.068938 \end{aligned}$$

Example II. — Let it be required to evaluate, by mechanical quadratures, the integrals

$$X_1 = \int_2^{3.2} 60\,T^3 dT \quad ; \quad X_2 = \int_2^{4.8} 60\,T^3 dT \quad ; \quad \text{and} \quad X_3 = \int_2^{11.6} 60\,T^3 dT$$

Here we tabulate ω times the given function for $T = 2, 4, 6, 8, 10, 12$; thus we obtain the following table of $F(T) \equiv 120\,T^3$:

T	$'F$	$F(T)\equiv 120T^3$	Δ'	Δ''	Δ'''
	$-$ 248				
2		960			
	$+$ 712		$+$ 6720		
4		7680		$+11520$	
	8392		18240		$+5760$
6		25920		17280	
	34312		35520		5760
8		61440		23040	
	95752		58560		$+5760$
10		120000		$+28800$	
	215752		$+87360$		
12		207360			
	$+423112$				

The several values of X here required are conveniently computed by the formulae (250). Thus (assuming $t=2$) the first step is to determine $'F_0$, the computation of which is as follows:

$$\begin{array}{llll} F_0 = +\ 960 & -\tfrac{1}{2}F_0 & = & -480 \\ \Delta'_0 = +\ 6720 & +\tfrac{1}{12}\Delta'_0 & = & +560 \\ \Delta''_0 = +11520 & -\tfrac{1}{24}\Delta''_0 & = & -480 \\ \Delta'''_0 = +\ 5760 & +\tfrac{19}{720}\Delta'''_0 & = & +152 \\ \hline & \therefore\ 'F_0 & = & -248 \end{array}$$

The column $'F$ is now completed by successive additions of the functions F, as shown in the table above.

(1) To find X_1: Here we have

$$n = \frac{3.2-2}{2} = 0.60$$

With this value of n we readily find $'F_n$, F_n, Δ'_n, Δ''_n and Δ'''_n by interpolation, employing NEWTON's Formula; whence X_1 is computed by formula (a) of (250). The results are given below:

$$\begin{array}{llll} & 'F_n & = & -\ \ 26.816 \\ F_n = +\ 3932.16 & +\tfrac{1}{2}F_n & = & +1966.080 \\ \Delta'_n = +12940.8 & -\tfrac{1}{12}\Delta'_n & = & -1078.400 \\ \Delta''_n = +14976.0 & +\tfrac{1}{24}\Delta''_n & = & +\ 624.000 \\ \Delta'''_n = +\ 5760.0 & -\tfrac{19}{720}\Delta'''_n & = & -\ 152.000 \\ \hline & \therefore\ X_1 & = & +1332.864 \end{array}$$

All of the quantities above are mathematically exact, and hence the result may be rigorously compared with the known value of the integral: thus, since

$$\int 60T^3dT = 15T^4$$

we have

$$X_1 = 15(3.2^4-2^4) = 1332.864$$

which is identical with the foregoing result.

(2) To find X_2: We use the same formula as before, the value of n in this case being

$$n = \frac{4.8-2}{2} = 1.40$$

or an interval of 0.40 counted forward from the quantities $'F_1$, F_1, Δ_1', Δ_1'', and Δ_1'''. Accordingly we find

$$\begin{array}{llll} & & 'F_n & = +2461.504 \\ F_n & = +13271.04 & +\tfrac{1}{2}F_n & = +6635.520 \\ \Delta'_n & = +24460.8 & -\tfrac{1}{12}\Delta'_n & = -2038.400 \\ \Delta''_n & = +19584.0 & +\tfrac{1}{24}\Delta''_n & = +\ 816.000 \\ \Delta'''_n & = +\ 5760.0 & -\tfrac{19}{720}\Delta'''_n & = -\ 152.000 \\ \hline & & \therefore X_2 & = +7722.624 \end{array}$$

This result is also mathematically exact, as may be easily verified.

(3) To find X_3: Since here the upper limit falls near the end of the given series, we employ formula (b) of (250). In this instance the value of n is—

$$n = \frac{11.6-2}{2} = 4.80 = 5 - 0.20$$

which corresponds to an interval of 0.20 counted *backwards* from the quantities having the subscript *five*. We therefore obtain

$$\begin{array}{lllll} n+1 = 6-0.20 & \ldots\ldots\ldots & & 'F_{n+1} & = +373075.264 \\ n = 5-0.20 & F_n & = +187307.52 & -\tfrac{1}{2}F_n & = -\ 93653.760 \\ n-1 = 4-0.20 & \Delta'_{n-1} & = +\ 81139.2 & -\tfrac{1}{12}\Delta'_{n-1} & = -\quad 6761.600 \\ n-2 = 3-0.20 & \Delta''_{n-2} & = +\ 27648.0 & -\tfrac{1}{24}\Delta''_{n-2} & = -\quad 1152.000 \\ n-3 = 2-0.20 & \Delta'''_{n-3} & = +\quad 5760.0 & -\tfrac{19}{720}\Delta'''_{n-3} & = -\quad 152.000 \\ \hline & & & \therefore X_3 & = +271355.904 \end{array}$$

which is mathematically exact.

72. *Quadrature as Based upon* STIRLING'S *Formula of Interpolation.*—The preceding formulae of quadrature obviously involve the same disadvantages as are inherent in NEWTON'S Formula of interpolation. We now proceed to integrate the expression for $F(t+n\omega)$ as given by STIRLING'S Formula, thus obtaining more convenient and accurate formulae than those already derived. For this purpose, let

the schedule of functions (including $'F$) and differences be taken as below :

T	$'F$	$F(T)$	Δ'	Δ''	Δ'''
$t-2\omega$		F_{-2}		Δ''_{-2}	
			$\Delta'_{-\frac{3}{2}}$		$\Delta'''_{-\frac{3}{2}}$
$t-\omega$		F_{-1}		Δ''_{-1}	
	$'F_{-\frac{1}{2}}$		$\Delta'_{-\frac{1}{2}}$		$\Delta'''_{-\frac{1}{2}}$
t		F_0		Δ''_0	
	$'F_{\frac{1}{2}}$		$\Delta'_{\frac{1}{2}}$		$\Delta'''_{\frac{1}{2}}$
$t+\omega$		F_1		Δ''_1	
	$'F_{\frac{3}{2}}$		$\Delta'_{\frac{3}{2}}$		$\Delta'''_{\frac{3}{2}}$
$t+2\omega$		F_2		Δ''_2	
.	.	.	.	.	.
.	.	.	.	.	.
$t+(i-1)\omega$		F_{i-1}		Δ''_{i-1}	
	$'F_{i-\frac{1}{2}}$		$\Delta'_{i-\frac{1}{2}}$		$\Delta'''_{i-\frac{1}{2}}$
$t+i\omega$		F_i		Δ''_i	
	$'F_{i+\frac{1}{2}}$		$\Delta'_{i+\frac{1}{2}}$		$\Delta'''_{i+\frac{1}{2}}$
$t+(i+1)\omega$		F_{i+1}		Δ''_{i+1}	
			$\Delta'_{i+\frac{3}{2}}$		$\Delta'''_{i+\frac{3}{2}}$
$t+(i+2)\omega$		F_{i+2}		Δ''_{i+2}	

Reverting now to (104), an inspection of this equation shows clearly the law of formation of the successive coefficients in the second member : hence, adding the term in Δ^{vi}, we have

$$\begin{aligned} F(t+n\omega) = {} & F_0 + n\left(\frac{\Delta'_{-\frac{1}{2}}+\Delta'_{\frac{1}{2}}}{2}\right) + \frac{n^2}{2}\Delta_0'' + \frac{n(n^2-1)}{6}\left(\frac{\Delta'''_{-\frac{1}{2}}+\Delta'''_{\frac{1}{2}}}{2}\right) + \frac{n^2(n^2-1)}{24}\Delta_0^{\mathrm{iv}} \\ & + \frac{n(n^2-1)(n^2-4)}{120}\left(\frac{\Delta^{\mathrm{v}}_{-\frac{1}{2}}+\Delta^{\mathrm{v}}_{\frac{1}{2}}}{2}\right) + \frac{n^2(n^2-1)(n^2-4)}{720}\Delta_0^{\mathrm{vi}} + \dots \end{aligned} \tag{251}$$

Multiplying by dn and integrating, we obtain

$$\begin{aligned} \int F(t+n\omega)\,dn = {} & nF_0 + \tfrac{n^2}{4}(\Delta'_{-\frac{1}{2}}+\Delta'_{\frac{1}{2}}) + \tfrac{n^3}{6}\Delta_0'' + \tfrac{1}{24}\left(\tfrac{n^4}{2}-n^2\right)(\Delta'''_{-\frac{1}{2}}+\Delta'''_{\frac{1}{2}}) \\ & + \tfrac{1}{24}\left(\tfrac{n^5}{5}-\tfrac{n^3}{3}\right)\Delta_0^{\mathrm{iv}} + \tfrac{1}{240}\left(\tfrac{n^6}{6}-\tfrac{5}{4}n^4+2n^2\right)(\Delta^{\mathrm{v}}_{-\frac{1}{2}}+\Delta^{\mathrm{v}}_{\frac{1}{2}}) \\ & + \tfrac{1}{720}\left(\tfrac{n^7}{7}-n^5+\tfrac{4}{3}n^3\right)\Delta_0^{\mathrm{vi}} + \dots + M \end{aligned} \tag{252}$$

M being the constant of integration. If this integral is now taken between the limits $n = -\frac{1}{2}$ and $n = +\frac{1}{2}$, the coefficients of Δ', Δ''', Δ^{v}, ... evidently vanish, and we find, therefore,

$$\int_{-\frac{1}{2}}^{+\frac{1}{2}} F(t+n\omega)\,dn = F_0 + \tfrac{1}{24}\Delta_0'' - \tfrac{17}{5760}\Delta_0^{\mathrm{iv}} + \tfrac{367}{967680}\Delta_0^{\mathrm{vi}} - \dots \tag{253}$$

In like manner, we derive

$$\left.\begin{aligned}
\int_{\frac{1}{2}}^{\frac{3}{2}} F(t+n\omega)\,dn &= F_1 + \tfrac{1}{24}\Delta_1'' - \tfrac{17}{5760}\Delta_1^{\text{iv}} + \tfrac{367}{967680}\Delta_1^{\text{vi}} - \dots \\
&\dots\dots\dots\dots \\
\int_{i-\frac{1}{2}}^{i+\frac{1}{2}} F(t+n\omega)\,dn &= F_i + \tfrac{1}{24}\Delta_i'' - \tfrac{17}{5760}\Delta_i^{\text{iv}} + \tfrac{367}{967680}\Delta_i^{\text{vi}} - \dots
\end{aligned}\right\} \qquad (254)$$

Whence, by summation, we obtain

$$\int_{-\frac{1}{2}}^{i+\frac{1}{2}} F(t+n\omega)\,dn = \sum_{r=0}^{r=i} F_r + \tfrac{1}{24}\sum_{r=0}^{r=i}\Delta_r'' - \tfrac{17}{5760}\sum_{r=0}^{r=i}\Delta_r^{\text{iv}} + \tfrac{367}{967680}\sum_{r=0}^{r=i}\Delta_r^{\text{vi}} - \dots \qquad (255)$$

Upon substituting the relations

$$\left.\begin{aligned}
\sum_{r=0}^{r=i} F_r &= F_0+F_1+F_2 + \dots + F_i = {}'F_{i+\frac{1}{2}} - {}'F_{-\frac{1}{2}} \\
\sum_{r=0}^{r=i} \Delta_r'' &= \Delta_0''+\Delta_1''+\Delta_2''+ \dots + \Delta_i'' = \Delta'_{i+\frac{1}{2}} - \Delta'_{-\frac{1}{2}} \\
\sum_{r=0}^{r=i} \Delta_r^{\text{iv}} &= \Delta_0^{\text{iv}}+\Delta_1^{\text{iv}}+\Delta_2^{\text{iv}}+ \dots + \Delta_i^{\text{iv}} = \Delta'''_{i+\frac{1}{2}} - \Delta'''_{-\frac{1}{2}} \\
&\dots\dots\dots\dots
\end{aligned}\right\} \qquad (256)$$

in formula (255), the latter becomes

$$\begin{aligned}
\int_{-\frac{1}{2}}^{i+\frac{1}{2}} F(t+n\omega)\,dn = ({}'F_{i+\frac{1}{2}} - {}'F_{-\frac{1}{2}}) + \tfrac{1}{24}(\Delta'_{i+\frac{1}{2}} - \Delta'_{-\frac{1}{2}}) - \tfrac{17}{5760}(\Delta'''_{i+\frac{1}{2}} - \Delta'''_{-\frac{1}{2}}) \\
+ \tfrac{367}{967680}(\Delta^{\text{v}}_{i+\frac{1}{2}} - \Delta^{\text{v}}_{-\frac{1}{2}}) - \dots
\end{aligned} \qquad (257)$$

Finally, therefore, we obtain

$$\int_{t-\frac{1}{2}\omega}^{t+i\omega+\frac{1}{2}\omega} F(T)\,dT = \omega \int_{-\frac{1}{2}}^{i+\frac{1}{2}} F(t+n\omega)\,dn$$

$$= \omega\left\{({}'F_{i+\frac{1}{2}} - {}'F_{-\frac{1}{2}}) + \tfrac{1}{24}(\Delta'_{i+\frac{1}{2}} - \Delta'_{-\frac{1}{2}}) - \tfrac{17}{5760}(\Delta'''_{i+\frac{1}{2}} - \Delta'''_{-\frac{1}{2}}) + \tfrac{367}{967680}(\Delta^{\text{v}}_{i+\frac{1}{2}} - \Delta^{\text{v}}_{-\frac{1}{2}}) - \dots\right\} \qquad (258)$$

When several values of an integral are to be computed from a given series, each having the lower limit $t-\frac{1}{2}\omega$, it will be more convenient and expeditious to determine the arbitrary quantity $'F_{-\frac{1}{2}}$ such that the sum of the terms with subscript $-\frac{1}{2}$ is equal to *zero*. The formula (258) may therefore be written as below:

$$\left.\begin{aligned}
{}'F_{-\frac{1}{2}} &= -\tfrac{1}{24}\Delta'_{-\frac{1}{2}} + \tfrac{17}{5760}\Delta'''_{-\frac{1}{2}} - \tfrac{367}{967680}\Delta^{\text{v}}_{-\frac{1}{2}} + \dots \\
\int_{t-\frac{1}{2}\omega}^{t+i\omega+\frac{1}{2}\omega} F(T)\,dT &= \omega \int_{-\frac{1}{2}}^{i+\frac{1}{2}} F(t+n\omega)\,dn \\
&= \omega({}'F_{i+\frac{1}{2}} + \tfrac{1}{24}\Delta'_{i+\frac{1}{2}} - \tfrac{17}{5760}\Delta'''_{i+\frac{1}{2}} + \tfrac{367}{967680}\Delta^{\text{v}}_{i+\frac{1}{2}} - \dots)
\end{aligned}\right\} \qquad (259)$$

As an application of (258), let it be required to find

$$X = \int_{30^\circ}^{45^\circ} \sec^2 T dT$$

Taking $\omega = 3^\circ$, $t = 31^\circ\ 30'$, and $'F_{-\frac{1}{2}} = 0$, we tabulate $F(T) \equiv \sec^2 T$ as follows:

T (° ′)	$'F$	$F(T) \equiv \sec^2 T$	Δ'	Δ''	Δ'''	Δ^{iv}
25 30		1.22751				
			+ 6729			
28 30		1.29480		+1343		
	0.00000		**8072**		+ **269**	
31 30		1.37552		1612		+ 78
	1.37552		9684		347	
34 30		1.47236		1959		117
	2.84788		11643		464	
37 30		1.58879		2423		155
	4.43667		14066		619	
40 30		1.72945		3042		223
	6.16612		17108		842	
43 30		1.90053		3884		+326
	8.06665		**20992**		**+1168**	
46 30		2.11045		+5052		
			+26044			
49 30		2.37089				

Owing to the rapid convergence of the coefficients in (258), the effect of fifth differences is here insensible: hence, using but three terms of this formula, we obtain

$(i = 4)$

$$\Delta'_{4\frac{1}{2}} - \Delta'_{-\frac{1}{2}} = +12920 \qquad\qquad 'F_{4\frac{1}{2}} - 'F_{-\frac{1}{2}} = +8.06665$$

$$\Delta'''_{4\frac{1}{2}} - \Delta'''_{-\frac{1}{2}} = +\ 899 \qquad\qquad +\tfrac{1}{24}(\Delta'_{4\frac{1}{2}} - \Delta'_{-\frac{1}{2}}) = +\ 538.3$$

$$-\tfrac{17}{5760}(\Delta'''_{4\frac{1}{2}} - \Delta'''_{-\frac{1}{2}}) = -\ 2.7$$

$$\log \Sigma = 0.906982 \qquad\qquad \Sigma = +8.07201$$

$$\log \omega = 8.718999 \qquad\qquad \omega = 3^\circ = \pi \div 60$$

$$\log X = 9.625981 \qquad\qquad \therefore X = 0.422650$$

Verification: Since

$$\int \sec^2 T dT = \tan T$$

we have

$$X = \tan 45^\circ - \tan 30^\circ = 1 - \tfrac{1}{3}\sqrt{3} = 0.422650$$

To illustrate the application of formula (259) when several values are assigned in succession to the integer i, we solve below an example which proceeds according to the evident relation

$$l = l_0 + \int_{T_0}^{T} \left(\frac{dl}{dT}\right) dT$$

where l denotes the value of any coördinate at the instant T, and l_0 its value at the epoch T_0. In particular, let us put

l = the heliocentric longitude of *Mars* for any date T;

$\frac{dl}{dT}$ = the daily motion in longitude;

T_0 = 1898 June 13, Greenwich mean noon;

l_0 = 1° 47′ 14″.3 = the heliocentric longitude for the date T_0;

and let it be required to compute the longitude (l) for Greenwich mean noon of the dates

(1898) June 21, June 29, July 7, July 15, and July 23;

the values of the daily motion being taken from the *American Ephemeris* for 1898.

The complete solution is conveniently arranged in tabular form as follows:

Date 1898	$F(T)=8\left(\frac{dl}{dT}\right)$	Δ'	Δ''	T	$l_0+{}'F$	$+\frac{\Delta'}{24}$	l
	° ′ ″	″	″		° ′ ″	″	° ′ ″
June 1	5 1 36.8						
		−105.5					
9	4 59 51.3		−20.4				
		125.9		June 13	1 47 19.5	−5.2	1 47 14
17	4 57 45.4		18.5				
		144.4		21	6 45 4.9	6.0	6 44 59
25	4 55 21.0		16.4				
		160.8		29	11 40 25.9	6.7	11 40 19
July 3	4 52 40.2		14.2				
		175.0		July 7	16 33 6.1	7.3	16 32 59
11	4 49 45.2		12.5				
		187.5		15	21 22 51.3	7.8	21 22 44
19	4 46 37.7		10.6				
		198.1		23	26 9 29.0	−8.3	26 9 21
27	4 43 19.6		− 8.2				
		−206.3					
Aug. 4	4 39 53.3						

The function tabulated in column $F(T)$ is *eight times* the daily motion in l: it is so multiplied, because the unit of the derivative being one day, we have $\omega = 8$; and thus the final multiplication by this factor is avoided.

Upon taking t = June 17, the formula (259) is at once applicable. We have, therefore, since differences beyond Δ'' are negligible,

$$'F_{-\frac{1}{2}} = -\tfrac{1}{24}\Delta'_{-\frac{1}{2}} = \frac{-125''.9}{-24} = +5''.2$$

and

$$l - l_0 = \int_{T_0=t-\frac{1}{2}\omega}^{T=t+(i+\frac{1}{2})\omega} \left(\frac{dl}{dT}\right) dT = {}'F_{i+\frac{1}{2}} + \tfrac{1}{24}\Delta'_{i+\frac{1}{2}}$$

the factor ω having been previously applied. Whence the expression for l becomes

$$l = l_0 + {}'F_{i+\frac{1}{2}} + \tfrac{1}{24}\,\Delta'_{i+\frac{1}{2}}$$

Thus, the value of l for *any* date T being found by adding the constant l_0 to the integral taken from T_0 to T, it is clear that we have merely to increase the above value of $'F_{-\frac{1}{2}}$ by the quantity $l_0 = 1^\circ\ 47'\ 14''.3$ in order to avoid the subsequent addition of this constant to each computed value of the integral. Accordingly, under the heading $l_0 + 'F$, on the line $t - \frac{1}{2}\omega$ (= June 13), we write the quantity $1^\circ\ 47'\ 19''.5$; the remaining numbers of this column are then formed in the usual manner by successive additions of the functions F. Each term of the series thus formed is evidently greater by l_0 than if the latter constant had been excluded from the initial term.

Under $+\frac{1}{24}\Delta'$ are written the values derived from the corresponding terms in Δ'. The sum $l_0 + 'F + \frac{1}{24}\Delta'$ is then tabulated in the final column, l, which therefore gives the heliocentric longitude of *Mars* for the dates indicated in column T.

73. *Applications in which the Limits Fall Otherwise than Midway Between Tabular Values of the Argument and Function.*—If we put

$$\theta(i+\tfrac{1}{2}) = {}'F_{i+\frac{1}{2}} + \tfrac{1}{24}\,\Delta'_{i+\frac{1}{2}} - \tfrac{17}{5760}\,\Delta'''_{i+\frac{1}{2}} + \tfrac{367}{967680}\,\Delta^{\mathrm{v}}_{i+\frac{1}{2}} - \cdot\ \cdot\ \cdot\ \cdot \tag{260}$$

the formula (257) becomes

$$\int_{-\frac{1}{2}}^{i+\frac{1}{2}} F(t+n\omega)\,dn = \theta(i+\tfrac{1}{2}) - \theta(-\tfrac{1}{2}) \tag{261}$$

Whence, if as before n denotes a fractional or mixed number, we derive, by the general method of interpolation employed in § 70,

$$\int_{-\frac{1}{2}}^{n} F(t+n\omega)\,dn = \theta(n) - \theta(-\tfrac{1}{2}) \tag{262}$$

Upon substituting n' and n'' successively for n in (262), and taking the difference of the resulting expressions, we get

$$\int_{n'}^{n''} F(t+n\omega)\,dn = \theta(n'') - \theta(n') \tag{263}$$

Finally, replacing the functions θ, in (262) and (263), according to the expression (260), we obtain the following formulae:

$$\int_{t-\frac{1}{2}\omega}^{t+n\omega} F(T)\,dT = \omega \int_{-\frac{1}{2}}^{n} F(t+n\omega)\,dn$$
$$= \omega \left\{ ('F_n - 'F_{-\frac{1}{2}}) + \tfrac{1}{24}(\Delta'_n - \Delta'_{-\frac{1}{2}}) - \tfrac{17}{5760}(\Delta'''_n - \Delta'''_{-\frac{1}{2}}) + \tfrac{367}{967680}(\Delta^{v}_n - \Delta^{v}_{-\frac{1}{2}}) - \ldots \right\} \qquad (264)$$

$$\int_{t+n'\omega}^{t+n''\omega} F(T)\,dT = \omega \int_{n'}^{n''} F(t+n\omega)\,dn$$
$$= \omega \left\{ ('F_{n''} - 'F_{n'}) + \tfrac{1}{24}(\Delta'_{n''} - \Delta'_{n'}) - \tfrac{17}{5760}(\Delta'''_{n''} - \Delta'''_{n'}) + \tfrac{367}{967680}(\Delta^{v}_{n''} - \Delta^{v}_{n'}) - \ldots \right\} \qquad (265)$$

where the quantity $'F_{-\frac{1}{2}}$ is wholly arbitrary; and where $'F_n$, Δ'_n, Δ'''_n, Δ^{v}_n, ... (and the similar terms with subscripts n' and n'') are to be found by interpolation.

When several values of an integral are to be computed from a given series by (264), the latter may be modified to the more expedient form given below:

$$\left. \begin{aligned} 'F_{-\frac{1}{2}} &= -\tfrac{1}{24}\Delta'_{-\frac{1}{2}} + \tfrac{17}{5760}\Delta'''_{-\frac{1}{2}} - \tfrac{367}{967680}\Delta^{v}_{-\frac{1}{2}} + \ldots \\ \int_{t-\frac{1}{2}\omega}^{t+n\omega} F(T)\,dT &= \omega \int_{-\frac{1}{2}}^{n} F(t+n\omega)\,dn \\ &= \omega\left('F_n + \tfrac{1}{24}\Delta'_n - \tfrac{17}{5760}\Delta'''_n + \tfrac{367}{967680}\Delta^{v}_n - \ldots\right) \end{aligned} \right\} \qquad (266)$$

EXAMPLE.—Find the value of

$$X = \int_{0.15}^{0.48} e^{T}\,dT$$

e being the base of the natural system of logarithms.

Taking $\omega = 0.1$, $t = 0.2$, and $F(T) \equiv e^{T}$, we prepare the following table:

T	$'F$	$F(T) \equiv e^{T}$	Δ'	Δ''	Δ'''	Δ^{iv}
0.0		1.000000				
			+105171			
0.1		1.105171		+11061		
	−0.004840		116232		+1163	
0.2		1.221403		12224		+123
	+1.216563		128456		1286	
0.3		1.349859		13510		134
	2.566422		141966		1420	
0.4		1.491825		14930		152
	4.058247		156896		1572	
0.5		1.648721		16502		+162
	+5.706968		173398		+1734	
0.6		1.822119		+18236		
			+191634			
0.7		2.013753				

Proceeding by formula (266), we find

$$'F_{-\frac{1}{2}} = 10^{-6}\left(-\tfrac{1}{24} \times 116232 + \tfrac{17}{5760} \times 1163\right) = -0.004840$$

whence the column $'F$ is completed as shown above. Denoting the assigned lower and upper limits by T' and T'', respectively, we have

$$T' = 0.15 = 0.20 - 0.05 = t - \tfrac{1}{4}\omega$$
$$T'' = 0.48 = 0.20 + 0.28 = t + 2.8\,\omega$$

Hence, at the upper limit, the value of n is—

$$n = 2.8 = 2.5 + 0.30$$

Accordingly, we find $'F_n$, Δ'_n, and Δ'''_n by interpolating forward from the quantities $'F_{2.5}$, $\Delta'_{2.5}$, and $\Delta'''_{2.5}$ with the interval 0.30. From the table above, we take

$$'F_{2.5} = +4.058247 \qquad \Delta'_{2.5} = +0.156896 \qquad \Delta'''_{2.5} = +0.001572$$

Hence, making the required interpolations by means of BESSEL'S Formula, and proceeding according to (266), we find

		$'F_n =$	$+4.535670.3$
$\Delta'_n =$	$+0.161674$	$+\frac{1}{24}\Delta'_n =$	$+\quad 6736.4$
$\Delta'''_n =$	$+\quad 1619$	$-\frac{17}{5760}\Delta'''_n =$	$-\quad 4.8$
		$\Sigma =$	$+4.542402$
		$\therefore X =$	$+0.4542402$

The true mathematical value of X is easily found: thus, since

$$\int e^T dT = e^T$$

we have

$$X = e^{0.48} - e^{0.15} = 0.454240159 \ldots .$$

74. *Quadrature as Based upon* BESSEL'S *Formula of Interpolation.*—Another set of formulae for mechanical quadrature, similar to those already developed, may be derived in the same manner from BESSEL'S expression for $F(t+n\omega)$. However, since these formulae may be obtained more conveniently by a direct transformation of those developed in the preceding section, we choose the latter course.

Putting $n'' = i$, and $n' = 0$, in formula (263), we have

$$\int_0^i F(t+n\omega)\,dn = \theta(i) - \theta(0) \tag{267}$$

We also have, by (260),

$$\theta(i) = {}'F_i + \tfrac{1}{24}\Delta'_i - \tfrac{17}{5760}\Delta'''_i + \tfrac{367}{967680}\Delta^{V}_i - \ldots . \tag{268}$$

Referring now to the general schedule on page 147, it will be observed that the quantities

$$'F_i, \ \Delta'_i, \ \Delta'''_i, \ \Delta^{v}_i, \ \ldots .$$

are not explicitly given, but must be found by interpolating *to halves* between $'F_{i-\frac{1}{2}}$ and $'F_{i+\frac{1}{2}}$, $\Delta'_{i-\frac{1}{2}}$ and $\Delta'_{i+\frac{1}{2}}$, etc., respectively. For this purpose, let us denote the algebraic *means* of the latter pairs of quantities by $('F_i)$, (Δ'_i), (Δ'''_i), ... ; that is, let us put

$$\left.\begin{aligned} ('F_i) &= \tfrac{1}{2}('F_{i-\frac{1}{2}} + 'F_{i+\frac{1}{2}}) \\ (\Delta'_i) &= \tfrac{1}{2}(\Delta'_{i-\frac{1}{2}} + \Delta'_{i+\frac{1}{2}}) \\ (\Delta'''_i) &= \tfrac{1}{2}(\Delta'''_{i-\frac{1}{2}} + \Delta'''_{i+\frac{1}{2}}) \\ &\ldots\ldots\ldots\ldots \end{aligned}\right\} \tag{269}$$

Applying formula (126), we have, therefore,

$$\left.\begin{aligned} 'F_i &= ('F_i) - \tfrac{1}{8}(\Delta'_i) + \tfrac{3}{128}(\Delta'''_i) - \tfrac{5}{1024}(\Delta^{v}_i)^* + \ldots . \\ \Delta'_i &= (\Delta'_i) - \tfrac{1}{8}(\Delta'''_i) + \tfrac{3}{128}(\Delta^{v}_i) - \ldots . \\ \Delta'''_i &= (\Delta'''_i) - \tfrac{1}{8}(\Delta^{v}_i) + \ldots . \\ \Delta^{v}_i &= (\Delta^{v}_i) - \ldots . \\ &\ldots\ldots\ldots\ldots\ldots\ldots \end{aligned}\right\} \tag{270}$$

Upon substituting these values of $'F_i$, Δ'_i, Δ'''_i, ... in (268), and reducing, we get

$$\theta(i) = ('F_i) - \tfrac{1}{12}(\Delta'_i) + \tfrac{11}{720}(\Delta'''_i) - \tfrac{191}{60480}(\Delta^{v}_i) + \ldots . \tag{271}$$

Putting $i = 0$, this becomes

$$\theta(0) = ('F_0) - \tfrac{1}{12}(\Delta'_0) + \tfrac{11}{720}(\Delta'''_0) - \tfrac{191}{60480}(\Delta^{v}_0) + \ldots . \tag{272}$$

Whence, from (267), we derive

$$\begin{aligned} \int_0^i F(t+n\omega)\,dn &= \theta(i) - \theta(0) \\ &= [('F_i) - ('F_0)] - \tfrac{1}{12}[(\Delta'_i) - (\Delta'_0)] \\ &\quad + \tfrac{11}{720}[(\Delta'''_i) - (\Delta'''_0)] - \tfrac{191}{60480}[(\Delta^{v}_i) - (\Delta^{v}_0)] + \ldots . \end{aligned} \tag{273}$$

---

* It is evident from (111) that the coefficient for the *sixth difference* in BESSEL'S Formula is—

$$\frac{(n+2)(n+1)\,n\,(n-1)(n-2)(n-3)}{\lfloor 6}$$

which, for $n = \frac{1}{2}$, yields the value given in the text.

Again, putting $n = i$ in (262), we have

$$\int_{-\frac{1}{2}}^{i} F(t+n\omega)\,dn = \theta(i) - \theta(-\tfrac{1}{2})$$
$$= ('F_i) - \tfrac{1}{12}(\Delta'_i) + \tfrac{11}{720}(\Delta'''_i) - \tfrac{191}{60480}(\Delta^{v}_i) + \cdots$$
$$-'F_{-\frac{1}{2}} - \tfrac{1}{24}\Delta'_{-\frac{1}{2}} + \tfrac{17}{5760}\Delta'''_{-\frac{1}{2}} - \tfrac{367}{967680}\Delta^{v}_{-\frac{1}{2}} + \cdots \quad (274)$$

In like manner, making $n'' = i + \frac{1}{2}$, and $n' = 0$, in (263), we obtain

$$\int_{0}^{i+\frac{1}{2}} F(t+n\omega)\,dn = \theta(i+\tfrac{1}{2}) - \theta(0)$$
$$= 'F_{i+\frac{1}{2}} + \tfrac{1}{24}\Delta'_{i+\frac{1}{2}} - \tfrac{17}{5760}\Delta'''_{i+\frac{1}{2}} + \tfrac{367}{967680}\Delta^{v}_{i+\frac{1}{2}} - \cdots$$
$$-('F_0) + \tfrac{1}{12}(\Delta'_0) - \tfrac{11}{720}(\Delta'''_0) + \tfrac{191}{60480}(\Delta^{v}_0) - \cdots \quad (275)$$

Finally, substituting $n'' = n$ and $n' = 0$, in (263), the latter becomes

$$\int_{0}^{n} F(t+n\omega)\,dn = \theta(n) - \theta(0)$$
$$= 'F_n + \tfrac{1}{24}\Delta'_n - \tfrac{17}{5760}\Delta'''_n + \tfrac{367}{967680}\Delta^{v}_n - \cdots$$
$$-('F_0) + \tfrac{1}{12}(\Delta'_0) - \tfrac{11}{720}(\Delta'''_0) + \tfrac{191}{60480}(\Delta^{v}_0) - \cdots \quad (276)$$

The equations (273), (274), (275) and (276) give, respectively, the following formulae of quadrature :

$$\int_{t}^{t+i\omega} F(T)\,dT = \omega\int_{0}^{i} F(t+n\omega)\,dn$$
$$= \omega\Big\{[('F_i)-('F_0)] - \tfrac{1}{12}[(\Delta'_i)-(\Delta'_0)] + \tfrac{11}{720}[(\Delta'''_i)-(\Delta'''_0)]$$
$$- \tfrac{191}{60480}[(\Delta^{v}_i)-(\Delta^{v}_0)] + \cdots\Big\} \quad (277)$$

$$\int_{t-\frac{1}{2}\omega}^{t+i\omega} F(T)\,dT = \omega\int_{-\frac{1}{2}}^{i} F(t+n\omega)\,dn$$
$$= \omega\Big\{('F_i) - \tfrac{1}{12}(\Delta'_i) + \tfrac{11}{720}(\Delta'''_i) - \tfrac{191}{60480}(\Delta^{v}_i) + \cdots$$
$$-'F_{-\frac{1}{2}} - \tfrac{1}{24}\Delta'_{-\frac{1}{2}} + \tfrac{17}{5760}\Delta'''_{-\frac{1}{2}} - \tfrac{367}{967680}\Delta^{v}_{-\frac{1}{2}} + \cdots\Big\} \quad (278)$$

$$\int_{t}^{t+i\omega+\frac{1}{2}\omega} F(T)\,dT = \omega\int_{0}^{i+\frac{1}{2}} F(t+n\omega)\,dn$$
$$= \omega\Big\{'F_{i+\frac{1}{2}} + \tfrac{1}{24}\Delta'_{i+\frac{1}{2}} - \tfrac{17}{5760}\Delta'''_{i+\frac{1}{2}} + \tfrac{367}{967680}\Delta^{v}_{i+\frac{1}{2}} - \cdots$$
$$-('F_0) + \tfrac{1}{12}(\Delta'_0) - \tfrac{11}{720}(\Delta'''_0) + \tfrac{191}{60480}(\Delta^{v}_0) - \cdots\Big\} \quad (279)$$

$$\int_{t}^{t+n\omega} F(T)\,dT = \omega\int_{0}^{n} F(t+n\omega)\,dn$$
$$= \omega\Big\{'F_n + \tfrac{1}{24}\Delta'_n - \tfrac{17}{5760}\Delta'''_n + \tfrac{367}{967680}\Delta^{v}_n - \cdots$$
$$-('F_0) + \tfrac{1}{12}(\Delta'_0) - \tfrac{11}{720}(\Delta'''_0) + \tfrac{191}{60480}(\Delta^{v}_0) - \cdots\Big\} \quad (280)$$

in which i denotes an integer and n a non-integer; where $'F_{-\frac{1}{2}}$ is wholly arbitrary; and where $('F_i)$, (Δ'_i), ... and $('F_0)$, (Δ'_0), ... are *means* of corresponding tabular quantities, as defined by (269).

If, in the formulae (277), (279), and (280), we take

$$('F_0) = \tfrac{1}{12}(\Delta'_0) - \tfrac{11}{720}(\Delta'''_0) + \tfrac{191}{60480}(\Delta^{v}_0) - \dots$$

then the sum of the terms with subscript *zero* will vanish. But, since

$$('F_0) = 'F_{-\frac{1}{2}} + \tfrac{1}{2}F_0$$

the preceding condition is evidently satisfied if we take

$$'F_{-\frac{1}{2}} = -\tfrac{1}{2}F_0 + \tfrac{1}{12}(\Delta'_0) - \tfrac{11}{720}(\Delta'''_0) + \tfrac{191}{60480}(\Delta^{v}_0) - \dots \qquad (281)$$

The formulae (277), (278), (279) and (280) may therefore be computed as follows:

$$\left.\begin{aligned} 'F_{-\frac{1}{2}} &= -\tfrac{1}{2}F_0 + \tfrac{1}{12}(\Delta'_0) - \tfrac{11}{720}(\Delta'''_0) + \tfrac{191}{60480}(\Delta^{v}_0) - \dots \\ \int_t^{t+i\omega} F(T)\,dT &= \omega\int_0^i F(t+n\omega)\,dn \\ &= \omega\left\{('F_i) - \tfrac{1}{12}(\Delta'_i) + \tfrac{11}{720}(\Delta'''_i) - \tfrac{191}{60480}(\Delta^{v}_i) + \dots\right\} \end{aligned}\right\} \qquad (282)$$

$$\left.\begin{aligned} 'F_{-\frac{1}{2}} &= -\tfrac{1}{24}\Delta'_{-\frac{1}{2}} + \tfrac{17}{5760}\Delta'''_{-\frac{1}{2}} - \tfrac{367}{967680}\Delta^{v}_{-\frac{1}{2}} + \dots \\ \int_{t-\frac{1}{2}\omega}^{t+i\omega} F(T)\,dT &= \omega\int_{-\frac{1}{2}}^{i} F(t+n\omega)\,dn \\ &= \omega\left\{('F_i) - \tfrac{1}{12}(\Delta'_i) + \tfrac{11}{720}(\Delta'''_i) - \tfrac{191}{60480}(\Delta^{v}_i) + \dots\right\} \end{aligned}\right\} \qquad (283)$$

$$\left.\begin{aligned} 'F_{-\frac{1}{2}} &= -\tfrac{1}{2}F_0 + \tfrac{1}{12}(\Delta'_0) - \tfrac{11}{720}(\Delta'''_0) + \tfrac{191}{60480}(\Delta^{v}_0) - \dots \\ \int_t^{t+i\omega+\frac{1}{2}\omega} F(T)\,dT &= \omega\int_0^{i+\frac{1}{2}} F(t+n\omega)\,dn \\ &= \omega\left('F_{i+\frac{1}{2}} + \tfrac{1}{24}\Delta'_{i+\frac{1}{2}} - \tfrac{17}{5760}\Delta'''_{i+\frac{1}{2}} + \tfrac{367}{967680}\Delta^{v}_{i+\frac{1}{2}} - \dots\right) \end{aligned}\right\} \qquad (284)$$

$$\left.\begin{aligned} 'F_{-\frac{1}{2}} &= -\tfrac{1}{2}F_0 + \tfrac{1}{12}(\Delta'_0) - \tfrac{11}{720}(\Delta'''_0) + \tfrac{191}{60480}(\Delta^{v}_0) - \dots \\ \int_t^{t+n\omega} F(T)\,dT &= \omega\int_0^{n} F(t+n\omega)\,dn \\ &= \omega\left('F_n + \tfrac{1}{24}\Delta'_n - \tfrac{17}{5760}\Delta'''_n + \tfrac{367}{967680}\Delta^{v}_0 - \dots\right) \end{aligned}\right\} \qquad (285)$$

Several examples will now be solved as an exercise in the use of the preceding formulae.

EXAMPLE I.—Let it be required to find

$$X = \int_0^{\frac{\pi}{2}} T \sin T\,dT$$

Here we take $\omega = 20° = \frac{\pi}{9}$, $t = 10° = \frac{\pi}{18}$, and tabulate $F(T) \equiv \omega T \sin T$, as follows:

T	$'F$	$F(T) \equiv \omega T \sin T$	Δ'	Δ''	Δ'''	Δ^{iv}	Δ^{v}
$-50°$		+0.23335					
			−14196				
30		0.09139		+ 6115			
			− 8081		+1966		
− 10		0.01058		8081		−1966	
	0.00000		0		0		0
+ 10		0.01058		8081		1966	
	0.01058		+ 8081		−1966		+361
30		0.09139		6115		1605	
	0.10197		14196		3571		648
50		0.23335		+ 2544		957	
	0.33532		16740		4528		856
70		0.40075		− 1984		− 101	
	0.73607		14756		4629		897
90		0.54831		6613		+ 796	
	1.28438		+ 8143		3833		+808
110		0.62974		10446		+1604	
			− 2303		−2229		
130		0.60671		−12675			
			−14978				
+150		+0.45693					

The value of X is now readily found by (278). Taking the arbitrary quantity $'F_{-\frac{1}{2}} = 0$, we complete the column $'F$ as above: we then have

$$'F_{-\frac{1}{2}} = \Delta'_{-\frac{1}{2}} = \Delta'''_{-\frac{1}{2}} = \Delta^{v}_{-\frac{1}{2}} = 0$$

Whence, proceeding by (278), we find

$$\begin{array}{llll} (i = 4) & & ('F_4) = \tfrac{1}{2}('F_{3\frac{1}{2}} + 'F_{4\frac{1}{2}}) & = +1.01022.5 \\ (\Delta'_4) = \tfrac{1}{2}(\Delta'_{3\frac{1}{2}} + \Delta'_{4\frac{1}{2}}) & = +0.11449.5 & -\tfrac{1}{12}(\Delta'_4) & = - \quad 954.1 \\ (\Delta'''_4) = \tfrac{1}{2}(\Delta'''_{3\frac{1}{2}} + \Delta'''_{4\frac{1}{2}}) & = - \quad 4231 & +\tfrac{11}{720}(\Delta'''_4) & = - \quad 64.6 \\ (\Delta^{v}_4) = \tfrac{1}{2}(\Delta^{v}_{3\frac{1}{2}} + \Delta^{v}_{4\frac{1}{2}}) & = + \quad 852 & -\tfrac{191}{60480}(\Delta^{v}_4) & = - \quad 2.7 \\ \hline & & \therefore X & = +1.00001 \end{array}$$

Verification: Since

$$\int T \sin T dT = \sin T - T \cos T$$

we have

$$X = \Big[\sin T - T \cos T\Big]_0^{\frac{\pi}{2}} = 1$$

Example II. — Compute the value of

$$X = \int_{0.9}^{1.2} \frac{dT}{(1 + 0.1\,T^2)^{\frac{1}{2}}}$$

Here we take $\omega = 0.1$, $t = 0.9$, and tabulate $F(T) \equiv (1 + 0.1\,T^2)^{-\frac{3}{2}}$, as below:

T	$'F$	$F(T) \equiv (1+0.1\,T^2)^{-\frac{3}{2}}$	Δ'	Δ''	Δ'''
0.7		0.93076	-1961		
0.8	-0.44672	0.91115	2141	-180	$+25$
0.9	$+0.44302$	0.88974	2296	155	27
1.0	1.30980	0.86678	2424	128	24
1.1	2.15234	0.84254	2528	104	25
1.2	$+2.96960$	0.81726	2607	79	$+22$
1.3		0.79119	-2664	-57	
1.4		0.76455			

Proceeding by means of (282), we compute $'F_{-\frac{1}{2}}$ as follows:

$$
\begin{array}{lll}
F_0 = +0.88974 & & -\tfrac{1}{2} F_0 = -0.44487 \\
(\Delta'_0) = - \quad 2218 & & +\tfrac{1}{12}(\Delta'_0) = - \quad 184.8 \\
(\Delta'''_0) = + \quad 26 & & -\tfrac{11}{720}(\Delta'''_0) = - \quad 0.4 \\
\hline
 & & \therefore\ 'F_{-\frac{1}{2}} = -0.44672
\end{array}
$$

Whence, having completed the column $'F$, we conclude the computation by (282), with the following results:

$$
\begin{array}{lll}
(i = 3) & & ('F_3) = +2.56097 \\
(\Delta'_3) = -0.02567.5 & & -\tfrac{1}{12}(\Delta'_3) = + \quad 214.0 \\
(\Delta'''_3) = + \quad 23.5 & & +\tfrac{11}{720}(\Delta'''_3) = + \quad 0.4 \\
\hline
 & & \Sigma = +2.56311 \\
 & & \therefore\ X = +0.256311
\end{array}
$$

Since

$$\int \frac{dT}{(1+0.1\,T^2)^{\frac{3}{2}}} = \frac{T}{(1+0.1\,T^2)^{\frac{1}{2}}}$$

we find for the true value of X,

$$X = 1.121936 - 0.865625 = 0.256311$$

Example III.—Let it be required to find

$$X = \int_{\frac{\pi}{4}}^{\tan^{-1}\frac{3}{2}} \sec^4 T\,dT$$

Expressing the assigned limits in degrees of arc, they become

$$\frac{\pi}{4} = 45^\circ \qquad \tan^{-1}\tfrac{3}{2} = 56^\circ\,18'\,35''.77 = 56^\circ.30994$$

We now take $\omega = 2° = \frac{\pi}{90}$, $t = 45°$, and tabulate the following values of $F(T) \equiv \omega \sec^4 T$:

T	$'F$	$F(T) \equiv \omega \sec^4 T$	Δ'	Δ''	Δ'''	Δ^{iv}	Δ^{v}
41°		0.10759					
			+ 1442				
43		0.12201		+ 320			
	−0.06819		1762		+ 90		
45		0.13963		410		+ 35	
	+0.07144		2172		125		+ 11
47		0.16135		535		46	
	0.23279		2707		171		20
49		0.18842		706		66	
	0.42121		3413		237		38
51		0.22255		943		104	
	0.64376		4356		341		51
53		0.26611		1284		155	
	0.90987		5640		496		86
55		0.32251		1780		241	
	1.23238		7420		737		+146
57		0.39671		2517		+387	
	+1.62909		9937		+1124		
59		0.49608		+3641			
			+13578				
61		0.63186					

Here we employ formula (285); in which, for the upper limit, we have

$$n = (56°.30994 - 45°) \div 2° = 5.65497 = 5.5 + 0.15497$$

For the value of $'F_{-\frac{1}{2}}$, we find

$$\begin{array}{llll} F_0 = +0.13963 & & -\tfrac{1}{2} F_0 & = -0.06981.5 \\ (\Delta'_0) = + \quad 1967 & & +\tfrac{1}{12} (\Delta'_0) & = + \quad 163.9 \\ (\Delta'''_0) = + \quad 108 & & -\tfrac{11}{720} (\Delta'''_0) & = - \quad 1.6 \\ \hline & & \therefore \ 'F_{-\frac{1}{2}} & = -0.06819 \end{array}$$

Whence, completing the series $'F$, and observing that the values of $'F_n$, Δ'_n, and Δ'''_n are obtained from their respective series by interpolation with the interval 0.15497, we find

$$\begin{array}{llll} & & 'F_n & = +1.28846.8 \\ \Delta'_n = +0.07754 & & +\tfrac{1}{24} \Delta'_n & = + \quad 323.1 \\ \Delta'''_n = + \quad 787 & & -\tfrac{17}{5760} \Delta'''_n & = - \quad 2.3 \\ \hline & & \therefore \ X & = +1.29168 \end{array}$$

Verification: The expression for the indefinite integral is—

$$\int \sec^4 T dT = \tan T + \tfrac{1}{3} \tan^3 T$$

Therefore

$$X = \left\{\tfrac{3}{2} + \tfrac{1}{3} \left(\tfrac{3}{2}\right)^3\right\} - \left\{1 + \tfrac{1}{3}\right\} = 1.29167$$

with which the above result substantially agrees.

Double Integration by Quadratures.

75. Having derived various formulae for the mechanical quadrature of single integrals, the corresponding formulae for *double integration* are now readily deduced. These will serve to compute integrals of the form

$$Y = \int\int_{T'}^{T''} F(T)\,dT^2 \tag{286}$$

independently of the analytical nature of the function $F(T)$, provided T' and T'' are numerically assigned. To define the quantity Y more explicitly, let us put

$$\int F(T)\,dT = f(T) + M \tag{286a}$$

where M is the constant of integration. We then have

$$Y = \int_{T'}^{T''} f(T)\,dT \quad + M(T'' - T') \tag{287}$$

It is therefore evident that unless the constant M has a definite value in any given case, the value of Y will be indeterminate. In practical applications, however, the quantity M is generally known from the fact that the *first integral* has an assigned value (usually *zero*) corresponding to the lower limit of integration.

If we now put

$$T = t + n\omega \quad , \quad T' = t + n'\omega \quad , \quad T'' = t + n''\omega$$

we have

$$dT^2 = \omega^2 dn^2 \tag{288}$$

and hence (286) becomes

$$Y = \int\int_{T'}^{T''} F(T)\,dT^2 = \omega^2 \int\int_{n'}^{n''} F(t + n\omega)\,dn^2 \tag{289}$$

upon which relation the subsequent formulae are based.

76. *Double Integration as Based upon* Newton's *Formula of Interpolation.*—If we substitute, successively, n' and n'' for n in (243), and take the difference of the two results, we obtain

$$\int_{n'}^{n''} F(t + n\omega)\,dn = \Psi(n'') - \Psi(n') \tag{290}$$

From the *form* of (290) it follows that the expression for the *indefinite* integral is—

$$\int F(t+n\omega)\,dn = \Psi(n)$$

or, by (238),

$$\int F(t+n\omega)\,dn = \int F_n\,dn = {}'F_n + \tfrac{1}{2} F_n + \beta\Delta'_n + \gamma\Delta''_n + \delta\Delta'''_n + \ldots \qquad (291)$$

the constant of integration being contained in $'F_n$, which depends upon the arbitrary quantity $'F_0$. Multiplying this equation by dn, and integrating, we get

$$\int\!\!\int F(t+n\omega)\,dn^2 = \int {}'F_n\,dn + \tfrac{1}{2}\int F_n\,dn + \beta\int \Delta'_n\,dn + \gamma\int \Delta''_n\,dn + \delta\int \Delta'''_n\,dn + \ldots \qquad (292)$$

Let us now consider a new series, namely—

$$''F_0,\ ''F_1,\ ''F_2,\ ''F_3,\ \ldots\ ''F_{i+2}$$

the term $''F_0$ being arbitrary, and the subsequent terms so determined that the quantities

$$'F_0,\ 'F_1,\ 'F_2,\ \ldots\ 'F_{i+1}$$

are the successive first differences of the proposed series. The manner of arranging the series $''F$, $'F$, and F, together with the differences of F, is shown in the schedule below:

T	$''F$	$'F$	$F(T)$	Δ'	Δ''	Δ'''	Δ^{iv}
	$''F_0$						
		$'F_0$					
t	$''F_1$		F_0				
		$'F_1$		Δ'_0			
$t+\omega$	$''F_2$		F_1		Δ''_0		
		$'F_2$		Δ'_1		Δ'''_0	
$t+2\omega$	$''F_3$		F_2		Δ''_1		Δ^{iv}_0
		$'F_3$		Δ'_2		Δ'''_1	
$t+3\omega$	$''F_4$		F_3		Δ''_2		Δ^{iv}_1
						.	
.							.
.							
.	.		.		.		.
		.		.		.	
$t+(i-2)\omega$	$''F_{i-1}$		F_{i-2}		Δ''_{i-3}		Δ^{iv}_{i-4}
		$'F_{i-1}$		Δ'_{i-2}		Δ'''_{i-3}	
$t+(i-1)\omega$	$''F_i$		F_{i-1}		Δ''_{i-2}		
		$'F_i$		Δ'_{i-1}			
$t+i\omega$	$''F_{i+1}$		F_i				
		$'F_{i+1}$					
	$''F_{i+2}$						

Now, since the differences $\Delta^{(r)}$ may be regarded as a series of functions whose 1st, 2d, ... differences are $\Delta^{(r+1)}$, $\Delta^{(r+2)}$ . . . . , it is clear that formula (291) may be applied successively to each of the integrals in the second member of (292). Accordingly, we have

$$\left.\begin{aligned}
\int {}'F_n\,dn &= {}''F_n + \tfrac{1}{2}\,{}'F_n + \beta F_n + \gamma\Delta'_n + \delta\Delta''_n + \epsilon\Delta'''_n + \ldots\ldots \\
\tfrac{1}{2}\int F_n\,dn &= \qquad \tfrac{1}{2}\left({}'F_n + \tfrac{1}{2}F_n + \beta\Delta'_n + \gamma\Delta''_n + \delta\Delta'''_n + \ldots\ldots\right) \\
\beta\int \Delta'_n\,dn &= \qquad\qquad \beta\left(F_n + \tfrac{1}{2}\Delta'_n + \beta\Delta''_n + \gamma\Delta'''_n + \ldots\ldots\right) \\
\gamma\int \Delta''_n\,dn &= \qquad\qquad\qquad \gamma\left(\Delta'_n + \tfrac{1}{2}\Delta''_n + \beta\Delta'''_n + \ldots\ldots\right) \\
\delta\int \Delta'''_n\,dn &= \qquad\qquad\qquad\qquad \delta\left(\Delta''_n + \tfrac{1}{2}\Delta'''_n + \ldots\ldots\right) \\
\epsilon\int \Delta^{\text{iv}}_n\,dn &= \qquad\qquad\qquad\qquad\qquad \epsilon\left(\Delta'''_n + \ldots\ldots\right) \\
&\ldots\ldots\ldots\ldots\ldots\ldots\ldots\ldots
\end{aligned}\right\} \tag{293}$$

Summing these expressions, we find, in accordance with (292),

$$\begin{aligned}\iint F(t+n\omega)\,dn^2 = {}&{}''F_n + {}'F_n + (\tfrac{1}{4}+2\beta)\,F_n + (\beta+2\gamma)\,\Delta'_n \\ &+ (\beta^2+\gamma+2\delta)\,\Delta''_n + (2\beta\gamma+\delta+2\epsilon)\,\Delta'''_n + \ldots\ldots\end{aligned} \tag{294}$$

Upon substituting the numerical values of β, γ, δ, ... from (222), formula (294) becomes

$$\iint F(t+n\omega)\,dn^2 = {}''F_n + {}'F_n + \tfrac{1}{12}\,F_n - \tfrac{1}{240}\,\Delta''_n + \tfrac{1}{240}\,\Delta'''_n - \ldots\ldots \tag{294a}$$

the coefficient of Δ'_n reducing to zero. We proceed to determine the expansion to which the coefficients of this formula belong. For brevity, let us write (294) in the form

$$\iint F(t+n\omega)\,dn^2 = {}''F_n + {}'F_n + aF_n + b\Delta'_n + c\Delta''_n + d\Delta'''_n + \ldots\ldots \tag{295}$$

Now, from (228), we have

$$\frac{1}{\log(1+x)} = x^{-1} + \tfrac{1}{2}\,x^0 + \beta x + \gamma x^2 + \delta x^3 + \ldots\ldots \tag{296}$$

Also, let us put

$$w \equiv x^{-2} + x^{-1} + ax^0 + bx + cx^2 + dx^3 + \ldots\ldots \tag{297}$$

in which the coefficients are taken as in (295). Whence, since the second member of (295) is the combined sum of the second members in (293), it is evident that (297) may be resolved, conversely, as follows :

$$
\begin{aligned}
w = x^{-2} + \tfrac{1}{2}x^{-1} + \beta x^0 + \gamma x + \delta x^2 + \ldots\ldots \\
+\tfrac{1}{2}(x^{-1} + \tfrac{1}{2}x^0 + \beta x + \gamma x^2 + \ldots\ldots) \\
+\beta(x^0 + \tfrac{1}{2}x + \beta x^2 + \ldots\ldots) \\
+\gamma(x + \tfrac{1}{2}x^2 + \ldots\ldots) \\
+\delta(x^2 + \ldots\ldots) \\
+ \ldots\ldots
\end{aligned}
$$

which may be written

$$
\begin{aligned}
w = {} & x^{-1}(x^{-1} + \tfrac{1}{2}x^0 + \beta x + \gamma x^2 + \delta x^3 + \ldots\ldots) \\
& + \tfrac{1}{2}x^0(x^{-1} + \tfrac{1}{2}x^0 + \beta x + \gamma x^2 + \delta x^3 + \ldots\ldots) \\
& + \beta x\ (x^{-1} + \tfrac{1}{2}x^0 + \beta x + \gamma x^2 + \delta x^3 + \ldots\ldots) \\
& + \gamma x^2\ (x^{-1} + \tfrac{1}{2}x^0 + \beta x + \gamma x^2 + \delta x^3 + \ldots\ldots) \\
& + \delta x^3\ (x^{-1} + \tfrac{1}{2}x^0 + \beta x + \gamma x^2 + \delta x^3 + \ldots\ldots) \\
& + \ldots\ldots\ldots\ldots\ldots\ldots\ldots\ldots \\
= {} & (x^{-1} + \tfrac{1}{2}x^0 + \beta x + \gamma x^2 + \ldots\ldots)(x^{-1} + \tfrac{1}{2}x^0 + \beta x + \gamma x^2 + \ldots\ldots) \\
= {} & (x^{-1} + \tfrac{1}{2}x^0 + \beta x + \gamma x^2 + \delta x^3 + \ldots\ldots)^2
\end{aligned}
$$

Therefore, by (296), we have

$$
\begin{aligned}
w &= \left\{ \log(1+x) \right\}^{-2} = \left(x - \frac{x^2}{2} + \frac{x^3}{3} - \frac{x^4}{4} + \frac{x^5}{5} - \ldots\ldots \right)^{-2} \\
&= x^{-2} + x^{-1} + \tfrac{1}{12}x^0 - \tfrac{1}{240}x^2 + \tfrac{1}{240}x^3 - \tfrac{221}{60480}x^4 + \tfrac{19}{6048}x^5 - \ldots\ldots \qquad (298)
\end{aligned}
$$

Comparing (297) and (298), it follows that the coefficients of the former, and hence, also, those of (295), are the coefficients in the expansion of $[\log(1+x)]^{-2}$, as developed in (298). Whence, introducing these values of $a, b, c, d, \ldots\ldots$ in (295), we obtain

$$\iint F(t+n\omega)\,dn^2 = {}''F_n + {}'F_n + \tfrac{1}{12}F_n - \tfrac{1}{240}\Delta''_n + \tfrac{1}{240}\Delta'''_n - \tfrac{221}{60480}\Delta^{\text{iv}}_n + \tfrac{19}{6048}\Delta^{\text{v}}_n - \ldots \qquad (299)$$

as was found directly — in part — in (294a).

Let us now put

$$
\begin{aligned}
\lambda(n) &= {}''F_n + {}'F_n + aF_n + b\Delta'_n + c\Delta''_n + d\Delta'''_n + e\Delta^{\text{iv}}_n + \ldots\ldots \\
&= {}''F_n + {}'F_n + \tfrac{1}{12}F_n + 0\Delta'_n - \tfrac{1}{240}\Delta''_n + \tfrac{1}{240}\Delta'''_n - \tfrac{221}{60480}\Delta^{\text{iv}}_n + \ldots\ldots \qquad (300)
\end{aligned}
$$

and (299) becomes

$$\iint F(t+n\omega)\,dn^2 = \lambda(n) \qquad (301)$$

Whence, if the integral be taken between the two fractional limits, n' and n'', we shall have

$$\int\int_{n'}^{n''} F(t+n\omega)\, dn^2 = \lambda(n'') - \lambda(n') \tag{302}$$

And if we make the upper limit an integer, say $n'' = i$, we have

$$\int\int_{n'}^{i} F(t+n\omega)\, dn^2 = \lambda(i) - \lambda(n') \tag{303}$$

The last formula involves the disadvantage of employing differences Δ_i', Δ_i'', Δ_i''', ... which are not given when the tabulation of $F(T)$ ends with the quantity F_i. To remedy this defect, we proceed as follows: Put

$$v = \lambda(i) = {}''F_i + {}'F_i + aF_i + b\Delta'_i + c\Delta_i'' + d\Delta_i''' + e\Delta_i^{\text{iv}} + \ldots \tag{304}$$

and substitute for ${}''F_i$, ${}'F_i$, F_i, Δ_i', Δ_i'', ... the expressions

$$\left.\begin{aligned}
{}''F_i &= {}''F_{i+2} - 2\,{}'F_{i+1} + F_i \\
{}'F_i &= {}'F_{i+1} - F_i \\
F_i &= F_i \\
\Delta_i' &= \Delta'_{i-1} + \Delta''_{i-2} + \Delta'''_{i-3} + \Delta^{\text{iv}}_{i-4} + \ldots \\
\Delta_i'' &= \Delta''_{i-2} + 2\Delta'''_{i-3} + 3\Delta^{\text{iv}}_{i-4} + \ldots \\
\Delta_i''' &= \Delta'''_{i-3} + 3\Delta^{\text{iv}}_{i-4} + \ldots \\
\Delta_i^{\text{iv}} &= \Delta^{\text{iv}}_{i-4} + \ldots \\
&\ldots\ldots\ldots\ldots\ldots
\end{aligned}\right\} \tag{305}$$

Whence the integral (303) may at once be expressed in terms of the *available* differences, Δ'_{i-1}, Δ''_{i-2}, Δ'''_{i-3}, ... However, to avoid direct substitution, let us put, as in (229),

$$x = \frac{u}{1-u} \tag{306}$$

and we shall have

$$\left.\begin{aligned}
x^{-2} &= u^{-2}(1-u)^2 = u^{-2} - 2u^{-1} + u^0 \\
x^{-1} &= u^{-1}(1-u) = u^{-1} - u^0 \\
x^0 &= u^0 \\
x &= u(1-u)^{-1} = u + u^2 + u^3 + u^4 + \ldots \\
x^2 &= u^2(1-u)^{-2} = u^2 + 2u^3 + 3u^4 + \ldots \\
x^3 &= u^3(1-u)^{-3} = u^3 + 3u^4 + \ldots \\
x^4 &= u^4(1-u)^{-4} = u^4 + \ldots \\
&\ldots\ldots\ldots\ldots\ldots
\end{aligned}\right\} \tag{307}$$

Again, from (297), we have

$$w = x^{-2} + x^{-1} + ax^0 + bx + cx^2 + dx^3 + ex^4 + \ldots \qquad (308)$$

Now, it is evident that if the expressions (307) be substituted in the second member of (308), the algebraic process will be identical in form with that of substituting the expressions (305) in (304). The w operation involves the quantities

$$w \;;\; x^{-2},\ x^{-1},\ x^0,\ x,\ x^2,\ x^3,\ \ldots \;;\; u^{-2},\ u^{-1},\ u^0,\ u,\ u^2,\ u^3,\ \ldots \;;$$

while the v operation involves, in exactly the same manner, the quantities

$$v \;;\; {}''F_i,\ {}'F_i,\ F_i,\ \Delta'_i,\ \Delta''_i,\ \Delta'''_i,\ \ldots \;;\; {}''F_{i+2},\ {}'F_{i+1},\ F_i,\ \Delta'_{i-1},\ \Delta''_{i-2},\ \Delta'''_{i-3}\ \ldots \;;$$

Hence, if we perform the w operation, the result for v is at once known. But the expression which results from substituting (307) in (308) is obtained with greater expedition by the following process: From (298), we have

$$w = \{\log(1+x)\}^{-2}$$

Whence, by (306), we find

$$w = \{-\log(1-u)\}^{-2} = \{\log(1-u)\}^{-2}$$

the expansion of which is immediately obtained by writing $-u$ for x in the second member of (297). Thus we find

$$w = u^{-2} - u^{-1} + au^0 - bu + cu^2 - du^3 + eu^4 - \ldots \qquad (309)$$

Therefore, according to the preceding reasoning, the expression for v is—

$$v = {}''F_{i+2} - {}'F_{i+1} + aF_i - b\Delta'_{i-1} + c\Delta''_{i-2} - d\Delta'''_{i-3} + e\Delta^{\text{iv}}_{i-4} - \ldots$$

Denoting this expression by $\pi(i)$, and restoring the numerical values of a, b, c, ... from (300), we have

$$\begin{aligned} v = \pi(i) &= {}''F_{i+2} - {}'F_{i+1} + aF_i - b\Delta'_{i-1} + c\Delta''_{i-2} - d\Delta'''_{i-3} + e\Delta^{\text{iv}}_{i-4} - \ldots \\ &= {}''F_{i+2} - {}'F_{i+1} + \tfrac{1}{12}F_i - \tfrac{1}{240}\Delta''_{i-2} - \tfrac{1}{240}\Delta'''_{i-3} - \tfrac{221}{60480}\Delta^{\text{iv}}_{i-4} - \ldots \qquad (310) \end{aligned}$$

Whence, by (304) and (310),

$$\lambda(i) = v = \pi(i)$$

and the formula (303) becomes, therefore,

$$\int\!\!\int_{n'}^{i} F(t+n\omega)\,dn^2 = \pi(i) - \lambda(n') \qquad (311)$$

In the formula just proved the quantity i denotes an integer. Now, by the general method of interpolation employed in §70, it is easily shown that (311) is true for non-integral values of i. Thus, writing n'' for i, this formula becomes

$$\int\!\!\int_{n'}^{n''} F(t+n\omega)\,dn^2 = \pi(n'') - \lambda(n') \qquad (312)$$

We now bring together equations (300), (310), (302), (312) and (289), in the order named; observing that in the first two of these we may write $''F_{n+1}$ for $''F_n + 'F_n$ and for $''F_{n+2} - 'F_{n+1}$, respectively. Thus we obtain the following group:

$$\left.\begin{aligned}
\lambda(n) &= {}''F_{n+1} + \tfrac{1}{12}F_n - \tfrac{1}{240}\Delta''_n + \tfrac{1}{240}\Delta'''_n - \tfrac{221}{60480}\Delta^{\text{iv}}_n + \tfrac{19}{6048}\Delta^{\text{v}}_n - \dots \\
\pi(n) &= {}''F_{n+1} + \tfrac{1}{12}F_n - \tfrac{1}{240}\Delta''_{n-2} - \tfrac{1}{240}\Delta'''_{n-3} - \tfrac{221}{60480}\Delta^{\text{iv}}_{n-4} - \tfrac{19}{6048}\Delta^{\text{v}}_{n-5} - \dots \\
&\int\!\!\int_{n'}^{n''} F(t+n\omega)\,dn^2 = \lambda(n'') - \lambda(n') \\
&\int\!\!\int_{n'}^{n''} F(t+n\omega)\,dn^2 = \pi(n'') - \lambda(n') \\
Y &= \int\!\!\int_{t+n'\omega}^{t+n''\omega} F(T)\,dT^2 = \omega^2 \int\!\!\int_{n'}^{n''} F(t+n\omega)\,dn^2
\end{aligned}\right\} \qquad (313)$$

From this group are immediately derived all of the formulae given in the following section.

77. We have already remarked that in the process of single integration the value of the definite integral is wholly independent of the absolute value of $'F_0$, which may therefore be assigned arbitrarily. Similarly, in double integration, the quantity $''F_0$ may be taken at pleasure, the integral being independent of its absolute value. Per contra, the *double* integral will evidently vary with the value assigned to $'F_0$. Hence, unless $'F_0$ is fixed by some special consideration, the value of the double integral is indeterminate — a conclusion already derived from (287).

Now, as was previously remarked, the value of the first integral corresponding to the lower limit is usually known in practical applications. We shall therefore denote by H_0 the value of $\int F(T)\,dT$ which results when t is substituted for T. Then, by (291), we have

$$\begin{aligned} H_0 &= \left[\int F(T)\,dT\right]_{T=t} = \omega\left[\int F(t+n\omega)\,dn\right]_{n=0} \\ &= \omega\,({}'F_0+\tfrac{1}{2}F_0+\beta\Delta'_0+\gamma\Delta''_0+\delta\Delta'''_0+\epsilon\Delta^{\text{iv}}_0+\;.\;.\;.\;.) \\ &= \omega\,({}'F_1-\tfrac{1}{2}F_0+\beta\Delta'_0+\gamma\Delta''_0+\delta\Delta'''_0+\epsilon\Delta^{\text{iv}}_0+\;.\;.\;.\;.) \end{aligned}$$

or, upon restoring the numerical values of $\beta,\ \gamma,\ \delta,\ .\;.\;.\;.$ from (222), and transposing,

$${}'F_1 = \frac{H_0}{\omega}+\tfrac{1}{2}F_0+\tfrac{1}{12}\Delta'_0-\tfrac{1}{24}\Delta''_0+\tfrac{19}{720}\Delta'''_0-\tfrac{3}{160}\Delta^{\text{iv}}_0+\tfrac{863}{60480}\Delta^{\text{v}}_0-\;.\;.\;.\;. \tag{314}$$

which determines ${}'F_1$, and hence, also, the double integral Y, provided H_0 is known. In practice the value of H_0 is frequently *zero*.

Using (314) in conjunction with the relations (313), we obtain the several groups of quadrature formulae given below:

$$\left.\begin{aligned} {}'F_1 &= \frac{H_0}{\omega}+\tfrac{1}{2}F_0+\tfrac{1}{12}\Delta'_0-\tfrac{1}{24}\Delta''_0+\tfrac{19}{720}\Delta'''_0-\tfrac{3}{160}\Delta^{\text{iv}}_0+\tfrac{863}{60480}\Delta^{\text{v}}_0-\;.\;.\;.\;. \\ \iint_t^{t+i\omega} F(T)\,dT^2 &= \omega^2\iint_0^i F(t+n\omega)\,dn^2 \\ &= \omega^2\{({}''F_{i+1}-{}''F_1)+\tfrac{1}{12}(F_i-F_0)-\tfrac{1}{240}(\Delta''_i-\Delta''_0)+\tfrac{1}{240}(\Delta'''_i-\Delta'''_0) \\ &\qquad -\tfrac{221}{60480}(\Delta^{\text{iv}}_i-\Delta^{\text{iv}}_0)+\tfrac{19}{6048}(\Delta^{\text{v}}_i-\Delta^{\text{v}}_0)-\;.\;.\;.\;.\} \end{aligned}\right\} \tag{315}$$

$$\left.\begin{aligned} {}'F_1 &= \frac{H_0}{\omega}+\tfrac{1}{2}F_0+\tfrac{1}{12}\Delta'_0-\tfrac{1}{24}\Delta''_0+\tfrac{19}{720}\Delta'''_0-\tfrac{3}{160}\Delta^{\text{iv}}_0+\tfrac{863}{60480}\Delta^{\text{v}}_0-\;.\;.\;.\;. \\ \iint_t^{t+n\omega} F(T)\,dT^2 &= \omega^2\iint_0^n F(t+n\omega)\,dn^2 \\ &= \omega^2\{({}''F_{n+1}-{}''F_1)+\tfrac{1}{12}(F_n-F_0)-\tfrac{1}{240}(\Delta''_n-\Delta''_0)+\tfrac{1}{240}(\Delta'''_n-\Delta'''_0) \\ &\qquad -\tfrac{221}{60480}(\Delta^{\text{iv}}_n-\Delta^{\text{iv}}_0)+\tfrac{19}{6048}(\Delta^{\text{v}}_n-\Delta^{\text{v}}_0)-\;.\;.\;.\;.\} \end{aligned}\right\} \tag{316}$$

$$\left.\begin{aligned} {}'F_1 &= \frac{H_0}{\omega}+\tfrac{1}{2}F_0+\tfrac{1}{12}\Delta'_0-\tfrac{1}{24}\Delta''_0+\tfrac{19}{720}\Delta'''_0-\tfrac{3}{160}\Delta^{\text{iv}}_0+\tfrac{863}{60480}\Delta^{\text{v}}_0-\;.\;.\;.\;. \\ \iint_{t+n\omega}^{t+i\omega} F(T)\,dT^2 &= \omega^2\iint_n^i F(t+n\omega)\,dn^2 \\ &= \omega^2\{({}''F_{i+1}-{}''F_{n+1})+\tfrac{1}{12}(F_i-F_n)-\tfrac{1}{240}(\Delta''_i-\Delta''_n)+\tfrac{1}{240}(\Delta'''_i-\Delta'''_n) \\ &\qquad -\tfrac{221}{60480}(\Delta^{\text{iv}}_i-\Delta^{\text{iv}}_n)+\tfrac{19}{6048}(\Delta^{\text{v}}_i-\Delta^{\text{v}}_n)-\;.\;.\;.\;.\} \end{aligned}\right\} \tag{317}$$

$$\left.\begin{aligned} {}'F_1 &= \frac{H_0}{\omega}+\tfrac{1}{2}F_0+\tfrac{1}{12}\Delta'_0-\tfrac{1}{24}\Delta''_0+\tfrac{19}{720}\Delta'''_0-\tfrac{3}{160}\Delta^{\text{iv}}_0+\tfrac{863}{60480}\Delta^{\text{v}}_0-\;.\;.\;.\;. \\ \iint_{t+n'\omega}^{t+n''\omega} F(T)\,dT^2 &= \omega^2\iint_{n'}^{n''} F(t+n\omega)\,dn^2 \\ &= \omega^2\{({}''F_{n''+1}-{}''F_{n'+1})+\tfrac{1}{12}(F_{n''}-F_{n'})-\tfrac{1}{240}(\Delta''_{n''}-\Delta''_{n'})+\tfrac{1}{240}(\Delta'''_{n''}-\Delta'''_{n'}) \\ &\qquad -\tfrac{221}{60480}(\Delta^{\text{iv}}_{n''}-\Delta^{\text{iv}}_{n'})+\tfrac{19}{6048}(\Delta^{\text{v}}_{n''}-\Delta^{\text{v}}_{n'})-\;.\;.\;.\;.\} \end{aligned}\right\} \tag{318}$$

The foregoing formulae are applicable when the upper limit falls near the *beginning* of the tabular series. When the upper limits falls at or near the *end* of the given series, the following formulae — likewise derived from (313) — may be employed:

$$\left.\begin{aligned}
{}'F_1 &= \frac{H_0}{\omega} + \tfrac{1}{2} F_0 + \tfrac{1}{12} \Delta'_0 - \tfrac{1}{24} \Delta''_0 + \tfrac{19}{720} \Delta'''_0 - \tfrac{3}{160} \Delta^{iv}_0 + \tfrac{863}{60480} \Delta^{v}_0 - \cdots \\
\iint_t^{t+i\omega} F(T)\,dT^2 &= \omega^2 \iint_0^i F(t+n\omega)\,dn^2 \\
&= \omega^2 \{ ({}''F_{i+1} - {}''F_1) + \tfrac{1}{12} (F_i - F_0) - \tfrac{1}{240} (\Delta''_{i-2} - \Delta''_0) - \tfrac{1}{240} (\Delta'''_{i-3} + \Delta'''_0) \\
&\qquad - \tfrac{221}{60480} (\Delta^{iv}_{i-4} - \Delta^{iv}_0) - \tfrac{19}{6048} (\Delta^{v}_{i-5} + \Delta^{v}_0) - \cdots \}
\end{aligned}\right\} \quad (319)$$

$$\left.\begin{aligned}
{}'F_1 &= \frac{H_0}{\omega} + \tfrac{1}{2} F_0 + \tfrac{1}{12} \Delta'_0 - \tfrac{1}{24} \Delta''_0 + \tfrac{19}{720} \Delta'''_0 - \tfrac{3}{160} \Delta^{iv}_0 + \tfrac{863}{60480} \Delta^{v}_0 - \cdots \\
\iint_t^{t+n\omega} F(T)\,dT^2 &= \omega^2 \iint_0^n F(t+n\omega)\,dn^2 \\
&= \omega^2 \{ ({}''F_{n+1} - {}''F_1) + \tfrac{1}{12} (F_n - F_0) - \tfrac{1}{240} (\Delta''_{n-2} - \Delta''_0) - \tfrac{1}{240} (\Delta'''_{n-3} + \Delta'''_0) \\
&\qquad - \tfrac{221}{60480} (\Delta^{iv}_{n-4} - \Delta^{iv}_0) - \tfrac{19}{6048} (\Delta^{v}_{n-5} + \Delta^{v}_0) - \cdots \}
\end{aligned}\right\} \quad (320)$$

$$\left.\begin{aligned}
{}'F_1 &= \frac{H_0}{\omega} + \tfrac{1}{2} F_0 + \tfrac{1}{12} \Delta'_0 - \tfrac{1}{24} \Delta''_0 + \tfrac{19}{720} \Delta'''_0 - \tfrac{3}{160} \Delta^{iv}_0 + \tfrac{863}{60480} \Delta^{v}_0 - \cdots \\
\iint_{t+n\omega}^{t+i\omega} F(T)\,dT^2 &= \omega^2 \iint_n^i F(t+n\omega)\,dn^2 \\
&= \omega^2 \{ ({}''F_{i+1} - {}''F_{n+1}) + \tfrac{1}{12} (F_i - F_n) - \tfrac{1}{240} (\Delta''_{i-2} - \Delta''_n) - \tfrac{1}{240} (\Delta'''_{i-3} + \Delta'''_n) \\
&\qquad - \tfrac{221}{60480} (\Delta^{iv}_{i-4} - \Delta^{iv}_n) - \tfrac{19}{6048} (\Delta^{v}_{i-5} + \Delta^{v}_n) - \cdots \}
\end{aligned}\right\} \quad (321)$$

$$\left.\begin{aligned}
{}'F_1 &= \frac{H_0}{\omega} + \tfrac{1}{2} F_0 + \tfrac{1}{12} \Delta'_0 - \tfrac{1}{24} \Delta''_0 + \tfrac{19}{720} \Delta'''_0 - \tfrac{3}{160} \Delta^{iv}_0 + \tfrac{863}{60480} \Delta^{v}_0 - \cdots \\
\iint_{t+n'\omega}^{t+n''\omega} F(T)\,dT^2 &= \omega^2 \iint_{n'}^{n''} F(t+n\omega)\,dn^2 \\
&= \omega^2 \{ ({}''F_{n''+1} - {}''F_{n'+1}) + \tfrac{1}{12} (F_{n''} - F_{n'}) - \tfrac{1}{240} (\Delta''_{n''-2} - \Delta''_{n'}) - \tfrac{1}{240} (\Delta'''_{n''-3} + \Delta'''_{n'}) \\
&\qquad - \tfrac{221}{60480} (\Delta^{iv}_{n''-4} - \Delta^{iv}_{n'}) - \tfrac{19}{6048} (\Delta^{v}_{n''-5} + \Delta^{v}_{n'}) - \cdots \}
\end{aligned}\right\} \quad (322)$$

In applications of all the preceding formulae, the value of $''F_1$ (or of $''F_0$ when employed) is wholly arbitrary, and therefore may be assigned at pleasure in every case. But when (315), (316), (319) and (320) are applicable, it is frequently convenient to determine $''F_1$ such that

$$-{}''F_1 - \tfrac{1}{12} F_0 + \tfrac{1}{240} \Delta''_0 - \tfrac{1}{240} \Delta'''_0 + \tfrac{221}{60480} \Delta^{iv}_0 - \tfrac{19}{6048} \Delta^{v}_0 + \cdots = 0$$

The formulae in question then take the form as follows:

$$\left.\begin{aligned}
{}'F_1 &= \frac{H_0}{\omega} + \tfrac{1}{2} F_0 + \tfrac{1}{12} \Delta'_0 - \tfrac{1}{24} \Delta''_0 + \tfrac{19}{720} \Delta'''_0 - \tfrac{3}{160} \Delta^{\text{iv}}_0 + \tfrac{863}{60480} \Delta^{\text{v}}_0 - \dots \\
{}''F_1 &= -\tfrac{1}{12} F_0 + \tfrac{1}{240} \Delta''_0 - \tfrac{1}{240} \Delta'''_0 + \tfrac{221}{60480} \Delta^{\text{iv}}_0 - \tfrac{19}{6048} \Delta^{\text{v}}_0 + \dots \\
\int\!\!\int_t^{t+i\omega} F(T)\, dT^2 &= \omega^2 \int\!\!\int_0^i F(t+n\omega)\, dn^2 \\
&= \omega^2 \left({}''F_{i+1} + \tfrac{1}{12} F_i - \tfrac{1}{240} \Delta''_i + \tfrac{1}{240} \Delta'''_i - \tfrac{221}{60480} \Delta^{\text{iv}}_i + \tfrac{19}{6048} \Delta^{\text{v}}_i - \dots \right)
\end{aligned}\right\} \quad (323)$$

$$\left.\begin{aligned}
{}'F_1 &= \frac{H_0}{\omega} + \tfrac{1}{2} F_0 + \tfrac{1}{12} \Delta'_0 - \tfrac{1}{24} \Delta''_0 + \tfrac{19}{720} \Delta'''_0 - \tfrac{3}{160} \Delta^{\text{iv}}_0 + \tfrac{863}{60480} \Delta^{\text{v}}_0 - \dots \\
{}''F_1 &= -\tfrac{1}{12} F_0 + \tfrac{1}{240} \Delta''_0 - \tfrac{1}{240} \Delta'''_0 + \tfrac{221}{60480} \Delta^{\text{iv}}_0 - \tfrac{19}{6048} \Delta^{\text{v}}_0 + \dots \\
\int\!\!\int_t^{t+n\omega} F(T)\, dT^2 &= \omega^2 \int\!\!\int_0^n F(t+n\omega)\, dn^2 \\
&= \omega^2 \left({}''F_{n+1} + \tfrac{1}{12} F_n - \tfrac{1}{240} \Delta''_n + \tfrac{1}{240} \Delta'''_n - \tfrac{221}{60480} \Delta^{\text{iv}}_n + \tfrac{19}{6048} \Delta^{\text{v}}_n - \dots \right)
\end{aligned}\right\} \quad (324)$$

$$\left.\begin{aligned}
{}'F_1 &= \frac{H_0}{\omega} + \tfrac{1}{2} F_0 + \tfrac{1}{12} \Delta'_0 - \tfrac{1}{24} \Delta''_0 + \tfrac{19}{720} \Delta'''_0 - \tfrac{3}{160} \Delta^{\text{iv}}_0 + \tfrac{863}{60480} \Delta^{\text{v}}_0 - \dots \\
{}''F_1 &= -\tfrac{1}{12} F_0 + \tfrac{1}{240} \Delta''_0 - \tfrac{1}{240} \Delta'''_0 + \tfrac{221}{60480} \Delta^{\text{iv}}_0 - \tfrac{19}{6048} \Delta^{\text{v}}_0 + \dots \\
\int\!\!\int_t^{t+i\omega} F(T)\, dT^2 &= \omega^2 \int\!\!\int_0^i F(t+n\omega)\, dn^2 \\
&= \omega^2 \left({}''F_{i+1} + \tfrac{1}{12} F_i - \tfrac{1}{240} \Delta''_{i-2} - \tfrac{1}{240} \Delta'''_{i-3} - \tfrac{221}{60480} \Delta^{\text{iv}}_{i-4} - \tfrac{19}{6048} \Delta^{\text{v}}_{i-5} - \dots \right)
\end{aligned}\right\} \quad (325)$$

$$\left.\begin{aligned}
{}'F_1 &= \frac{H_0}{\omega} + \tfrac{1}{2} F_0 + \tfrac{1}{12} \Delta'_0 - \tfrac{1}{24} \Delta''_0 + \tfrac{19}{720} \Delta'''_0 - \tfrac{3}{160} \Delta^{\text{iv}}_0 + \tfrac{863}{60480} \Delta^{\text{v}}_0 - \dots \\
{}''F_1 &= -\tfrac{1}{12} F_0 + \tfrac{1}{240} \Delta''_0 - \tfrac{1}{240} \Delta'''_0 + \tfrac{221}{60480} \Delta^{\text{iv}}_0 - \tfrac{19}{6048} \Delta^{\text{v}}_0 + \dots \\
\int\!\!\int_t^{t+n\omega} F(T)\, dT^2 &= \omega^2 \int\!\!\int_0^n F(t+n\omega)\, dn^2 \\
&= \omega^2 \left({}''F_{n+1} + \tfrac{1}{12} F_n - \tfrac{1}{240} \Delta''_{n-2} - \tfrac{1}{240} \Delta'''_{n-3} - \tfrac{221}{60480} \Delta^{\text{iv}}_{n-4} - \tfrac{19}{6048} \Delta^{\text{v}}_{n-5} - \dots \right)
\end{aligned}\right\} \quad (326)$$

The differences which appear in the foregoing formulae, together with the auxiliary functions $'F$ and $''F$, are to be taken according to the schedule on page 161. The symbol i denotes a positive integer, while n designates a fractional or mixed number: so that all functions and differences whose subscripts involve n must be derived from their respective series by *interpolation*. Finally, the quantity H_0 denotes — as previously defined — the value of $\int F(T)\,dT$ when t is substituted for T: so that we have

$$H_0 = \left[\int F(T)\, dT\right]_{T=t} \quad (327)$$

It may happen occasionally that the value of H_0 is unknown, while the value of $\int F(T)\,dT$ corresponding to $T = t + n\omega$ is known for a particular value of n. Denoting this quantity by H_n, we may, by any one of the foregoing methods, compute the definite integral

$$X = \int_t^{t+n\omega} F(T)\,dT = H_n - H_0$$

and hence find

$$H_0 = H_n - X \qquad (327a)$$

with which value we proceed as before.

Several examples will now be solved as an exercise to illustrate the formulae given above.

Example I.—Let it be required to find

$$Y = \int\int_0^{\frac{\pi}{2}} \cos T dT^2$$

on the supposition that $\int \cos T dT = 2$ when $T = 0$.

We tabulate and difference the following values of $F(T) \equiv \cos T$:

T	$''F$	$'F$	$F(T) \equiv \cos T$	Δ'	Δ''	Δ'''	Δ^{iv}
0°	0.00000		1.00000				
		11.95916		− 1519			
10	11.95916		0.98481		−2993		
		12.94397		4512		+139	
20	24.90313		0.93969		2854		+82
		13.88366		7366		221	
30	38.78679		0.86603		2633		86
		14.74969		9999		307	
40	53.53648		0.76604		2326		65
		15.51573		12325		372	
50	69.05221		0.64279		1954		63
		16.15852		14279		435	
60	85.21073		0.50000		1519		45
		16.65852		15798		480	
70	101.86925		0.34202		1039		+31
		17.00054		16837		+511	
80	118.86979		0.17365		− 528		
		17.17419		−17365			
90	136.04398		0.00000				

Accordingly, we have

$$t = 0° \qquad \omega = 10° = \frac{\pi}{18} \qquad H_0 = 2 \qquad i = 9$$

Proceeding by (319), the computation of $'F_1$ is as follows:

$$
\begin{array}{llll}
 & & H_0 \div \omega & = +11.45915.6 \\
F_0 = +1.00000 & & +\tfrac{1}{2}\,F_0 & = +\ 0.50000.0 \\
\Delta'_0 = -\ \ 1519 & & +\tfrac{1}{12}\,\Delta'_0 & = -\qquad 126.6 \\
\Delta''_0 = -\ \ 2993 & & -\tfrac{1}{24}\,\Delta''_0 & = +\qquad 124.7 \\
\Delta'''_0 = +\ \ \ 139 & & +\tfrac{19}{720}\,\Delta'''_0 & = +\qquad\ \ 3.7 \\
\Delta^{iv}_0 = +\ \ \ \ 82 & & -\tfrac{3}{160}\,\Delta^{iv}_0 & = -\qquad\ \ 1.5 \\
\hline
 & & \therefore\ 'F_1 & = +11.95916
\end{array}
$$

The column $'F$ is now completed by successive additions; hence, also, the column $''F$, having first assumed $''F_1 = 0$. Whence, by (319), the remainder of the computation is as follows:

$$\begin{array}{llll} ''F_{10} = +136.04398 & ''F_1 = \quad 0.00000 & (''F_{10} - ''F_1) & = +136.04398 \\ F_9 = \quad 0.00000 & F_0 = +1.00000 & +\frac{1}{12}\,(F_9 - F_0) & = - \quad 0.08333.3 \\ \Delta_7'' = - \quad 528 & \Delta_0'' = - \quad 2993 & -\frac{1}{240}\,(\Delta_7'' - \Delta_0'') & = - \quad 10.3 \\ \Delta_6''' = + \quad 511 & \Delta_0''' = + \quad 139 & -\frac{1}{240}\,(\Delta_6''' + \Delta_0''') & = - \quad 2.7 \\ \Delta_5^{iv} = + \quad 31 & \Delta_0^{iv} = + \quad 82 & -\frac{221}{60480}\,(\Delta_5^{iv} - \Delta_0^{iv}) & = + \quad 0.2 \\ \hline & \log \Sigma = 2.1334129 & & \Sigma = +135.96052 \\ & \log \omega^2 = 8.4837548 & & \\ & \log Y = 0.6171677 & & \therefore\ Y = \quad 4.141595 \end{array}$$

To verify this result, we have

$$\int \cos T dT = \sin T + C$$

$$Y = \int\int_0^{\frac{\pi}{2}} \cos T dT^2 = \Big[-\cos T + CT\Big]_0^{\frac{\pi}{2}} = 1 + \tfrac{1}{2} C\pi$$

where C is the constant of the first integration. To determine C, the first of these relations gives

$$H_0 = \Big[\sin T + C\Big]_{T=0} = C$$

whence

$$C = 2$$

and therefore

$$Y = 1 + \pi = 4.141593$$

EXAMPLE II.—Compute the value of

$$Y = \int\int_2^{2.468} T^{-2} dT^2$$

which corresponds to $H_0 = 0$.

Here we tabulate and difference $F(T) \equiv T^{-2}$ as below:

T	$''F$	$'F$	$F(T) \equiv T^{-2}$	Δ'	Δ''	Δ'''
2.0	−0.02082		0.25000			
		+0.12292		−2324		
2.1	+0.10210		0.22676		+309	
		0.34968		2015		−51
2.2	0.45178		0.20661		258	
		0.55629		1757		44
2.3	1.00807		0.18904		214	
		0.71533		1543		−32
2.4	1.75340		0.17361		+182	
		+0.91894		−1361		
2.5	+2.67234		0.16000			

We have, therefore,

$$t = 2.0 \qquad \omega = 0.1 \qquad H_0 = 0$$

whence, proceeding by (326), the computation of $'F_1$ and $''F_1$ is as follows:

$$\begin{array}{rl} +\frac{1}{2}F_0 = & +0.12500 \\ +\frac{1}{12}\Delta'_0 = & -\quad 193.7 \\ -\frac{1}{24}\Delta''_0 = & -\quad 12.9 \\ +\frac{19}{720}\Delta'''_0 = & -\quad 1.3 \\ \hline \therefore\ 'F_1 = & +0.12292 \end{array} \qquad \begin{array}{rl} \ldots\ldots & \\ -\frac{1}{12}F_0 = & -0.02083.3 \\ +\frac{1}{240}\Delta''_0 = & +\quad 1.3 \\ -\frac{1}{240}\Delta'''_0 = & +\quad 0.2 \\ \hline \therefore\ ''F_1 = & -0.02082 \end{array}$$

From the completed table we now find

$$\begin{array}{rl} n = & (2.468-2.0)\div 0.1 \\ = & 4.68 = 5 - 0.32 \\ F_n = & +0.16418 \\ \Delta''_{n-2} = & +\quad 191 \\ \Delta'''_{n-3} = & -\quad 36 \end{array} \qquad \begin{array}{rl} \ldots\ldots & \\ ''F_{n+1} = & +2.36025.6 \\ +\frac{1}{12}F_n = & +\quad 1368.2 \\ -\frac{1}{240}\Delta''_{n-2} = & -\quad 0.8 \\ -\frac{1}{240}\Delta'''_{n-3} = & +\quad 0.1 \\ \hline \Sigma = & +2.37393 \\ \therefore\ Y = & +0.0237393 \end{array}$$

This result is easily verified, for we have

$$\int T^{-2}dT = -\frac{1}{T} + C$$

$$Y = \Big[-\log_e T + CT\Big]_2^{2.468} = -\log_e 1.234 + 0.468\,C$$

also

$$0 = H_0 = \Big[-\frac{1}{T} + C\Big]_{T=2} = -\tfrac{1}{2} + C$$

$$\therefore\ C = \tfrac{1}{2}$$

Hence

$$Y = -\log_e 1.234 + 0.234 = -0.2102609 + 0.234 = +0.0237391$$

with which the above result substantially agrees.

EXAMPLE III.—From the table of the preceding example, find the value of

$$Y = \int\!\!\int_2^{2.15} T^{-2}dT^2$$

Here we employ formula (324), in which we take

$$n = \frac{2.15 - 2.0}{0.1} = 1.50 = 1 + \tfrac{1}{2}$$

We therefore obtain

$(n + 1 = 2 + \frac{1}{2})$		$''F_{n+1}$	$= +0.24992.0$
F_n	$= +0.21633$	$+\frac{1}{12} F_n$	$= + \quad 1802.8$
Δ''_n	$= + \quad 235$	$-\frac{1}{240} \Delta''_n$	$= - \quad 1.0$
Δ'''_n	$= - \quad 38$	$+\frac{1}{240} \Delta'''_n$	$= - \quad 0.2$
		Σ	$= +0.26794$
		$\therefore Y$	$= +0.0026794$

The true mathematical value of Y is—

$$Y = 0.075 - \log_e 1.075 = +0.0026793 \ldots.$$

78. *Double Integration as Based upon* STIRLING'S *and* BESSEL'S *Formulae of Interpolation.*—Let the schedule of functions (including $'F$ and $''F$) and differences to be used in the subsequent formulae of quadrature be as follows:

T	$''F$	$'F$	$F(T)$	Δ'	Δ''	Δ'''
$t - 2\omega$			F_{-2}		Δ''_{-2}	
				$\Delta'_{-\frac{3}{2}}$		$\Delta'''_{-\frac{3}{2}}$
$t - \omega$	$''F_{-1}$		F_{-1}		Δ''_{-1}	
		$'F_{-\frac{1}{2}}$		$\Delta'_{-\frac{1}{2}}$		$\Delta'''_{-\frac{1}{2}}$
t	$''F_0$		F_0		Δ''_0	
		$'F_{\frac{1}{2}}$		$\Delta'_{\frac{1}{2}}$		$\Delta'''_{\frac{1}{2}}$
$t + \omega$	$''F_1$		F_1		Δ''_1	
		$'F_{\frac{3}{2}}$		$\Delta'_{\frac{3}{2}}$		$\Delta'''_{\frac{3}{2}}$
$t + 2\omega$	$''F_2$		F_2		Δ''_2	
.	.	.	.	.	.	.
.		.				
.	.	.	.	.	.	.
$t + (i-1)\omega$	$''F_{i-1}$		F_{i-1}		Δ''_{i-1}	
		$'F_{i-\frac{1}{2}}$		$\Delta'_{i-\frac{1}{2}}$		$\Delta'''_{i-\frac{1}{2}}$
$t + i\omega$	$''F_i$		F_i		Δ''_i	
		$'F_{i+\frac{1}{2}}$		$\Delta'_{i+\frac{1}{2}}$		$\Delta'''_{i+\frac{1}{2}}$
$t + (i+1)\omega$	$''F_{i+1}$		F_{i+1}		Δ''_{i+1}	
				$\Delta'_{i+\frac{3}{2}}$		$\Delta'''_{i+\frac{3}{2}}$
$t + (i+2)\omega$			F_{i+2}		Δ''_{i+2}	

From the form of (263) it follows that the expression for the indefinite integral of $F(t + n\omega)\,dn$ is—

$$\int F(t + n\omega)\,dn = \theta(n) \tag{328}$$

Now, by (260), we have

$$\theta(n) = {}'F_n + \tfrac{1}{24}\Delta'_n - \tfrac{17}{5760}\Delta'''_n + \tfrac{367}{967680}\Delta^{v}_n - \ldots .$$

and hence the preceding equation becomes

$$\int F(t+n\omega)\,dn = {}'F_n + \tfrac{1}{24}\Delta'_n - \tfrac{17}{5760}\Delta'''_n + \tfrac{367}{967680}\Delta^{v}_n - \ldots . \tag{328a}$$

For brevity, let us put

$$a = +\tfrac{1}{24}. \qquad b = -\tfrac{17}{5760} \qquad c = +\tfrac{367}{967680} \tag{329}$$

and (328a) may be written

$$\int F(t+n\omega)\,dn = \int F_n\,dn = {}'F_n + a\Delta'_n + b\Delta'''_n + c\Delta^{v}_n + \ldots . \tag{330}$$

the constant of integration being contained in $'F_n$. Multiplying this equation by dn, and integrating, we get

$$\int\!\!\int F(t+n\omega)\,dn^2 = \int {}'F_n\,dn + a\int \Delta'_n\,dn + b\int \Delta'''_n\,dn + c\int \Delta^{v}_n\,dn + \ldots . \tag{331}$$

Applying formula (330) successively to each of the integrals expressed in the second member of (331), we obtain

$$\begin{aligned}\int\!\!\int F(t+n\omega)\,dn^2 = {}''F_n + aF_n &+ b\Delta''_n + c\Delta^{iv}_n + \ldots .\\ +a(F_n + a\Delta''_n &+ b\Delta^{iv}_n + \ldots .)\\ +b(\Delta''_n &+ a\Delta^{iv}_n + \ldots .)\\ &+c(\Delta^{iv}_n + \ldots .)\\ &\qquad + \ldots .\\ = {}''F_n + 2aF_n &+ (a^2+2b)\,\Delta''_n + 2\,(ab+c)\,\Delta^{iv}_n + \ldots .\end{aligned}$$

Whence, restoring the values of a, b, c, ... from (329), and reducing, we obtain

$$\int\!\!\int F(t+n\omega)\,dn^2 = {}''F_n + \tfrac{1}{12}F_n - \tfrac{1}{240}\Delta''_n + \tfrac{31}{60480}\Delta^{iv}_n - \ldots . \tag{332}$$

If, as in (327), we denote by H_0 the value of $\int F(T)\,dT$ which obtains for $T = t$, then, by (328), we have

$$H_0 = \left[\int F(T)\,dT\right]_{T=t} = \omega\left[\int F(t+n\omega)\,dn\right]_{n=0} = \omega.\theta(0)$$

and hence, by (272),

$$H_0 = \omega\{({}'F_0) - \tfrac{1}{12}(\Delta'_0) + \tfrac{11}{720}(\Delta'''_0) - \tfrac{191}{60480}(\Delta^{v}_0) + \ldots .\} \tag{333}$$

Upon substituting $i=0$ in the first of equations (269), we get

$$('F_0) = \tfrac{1}{2}('F_{-\frac{1}{2}}+'F_{\frac{1}{2}}) = 'F_{\frac{1}{2}} - \tfrac{1}{2}F_0$$

which, together with (333), gives

$$'F_{\frac{1}{2}} = \frac{H_0}{\omega} + \tfrac{1}{2}F_0 + \tfrac{1}{12}(\Delta'_0) - \tfrac{11}{720}(\Delta'''_0) + \tfrac{191}{60480}(\Delta^{\mathrm{v}}_0) - \cdot\;\cdot\;\cdot\;\cdot \tag{334}$$

where the differences enclosed within parentheses are *means* of the corresponding tabular quantities, as defined by (269).

By employing simultaneously the relations (332) and (334), and assigning various limits to the integral, we obtain the following group of formulae :

$$\left.\begin{aligned}
'F_{\frac{1}{2}} &= \frac{H_0}{\omega} + \tfrac{1}{2}F_0 + \tfrac{1}{12}(\Delta'_0) - \tfrac{11}{720}(\Delta'''_0) + \tfrac{191}{60480}(\Delta^{\mathrm{v}}_0) - \cdot\;\cdot\;\cdot\;\cdot \\
\int\!\!\int_t^{t+i\omega} F(T)\,dT^2 &= \omega^2\int\!\!\int_0^i F(t+n\omega)\,dn^2 \\
&= \omega^2\{(''F_i - ''F_0) + \tfrac{1}{12}(F_i - F_0) - \tfrac{1}{240}(\Delta''_i - \Delta''_0) + \tfrac{31}{60480}(\Delta^{\mathrm{iv}}_i - \Delta^{\mathrm{iv}}_0) - \cdot\;\cdot\;\cdot\}
\end{aligned}\right\} \tag{335}$$

$$\left.\begin{aligned}
'F_{\frac{1}{2}} &= \frac{H_0}{\omega} + \tfrac{1}{2}F_0 + \tfrac{1}{12}(\Delta'_0) - \tfrac{11}{720}(\Delta'''_0) + \tfrac{191}{60480}(\Delta^{\mathrm{v}}_0) - \cdot\;\cdot\;\cdot\;\cdot \\
\int\!\!\int_t^{t+n\omega} F(T)\,dT^2 &= \omega^2\int\!\!\int_0^n F(t+n\omega)\,dn^2 \\
&= \omega^2\{(''F_n - ''F_0) + \tfrac{1}{12}(F_n - F_0) - \tfrac{1}{240}(\Delta''_n - \Delta''_0) + \tfrac{31}{60480}(\Delta^{\mathrm{iv}}_n - \Delta^{\mathrm{iv}}_0) - \cdot\;\cdot\;\cdot\}
\end{aligned}\right\} \tag{336}$$

$$\left.\begin{aligned}
'F_{\frac{1}{2}} &= \frac{H_0}{\omega} + \tfrac{1}{2}F_0 + \tfrac{1}{12}(\Delta'_0) - \tfrac{11}{720}(\Delta'''_0) + \tfrac{191}{60480}(\Delta^{\mathrm{v}}_0) - \cdot\;\cdot\;\cdot\;\cdot \\
\int\!\!\int_{t+n\omega}^{t+i\omega} F(T)\,dT^2 &= \omega^2\int\!\!\int_n^i F(t+n\omega)\,dn^2 \\
&= \omega^2\{(''F_i - ''F_n) + \tfrac{1}{12}(F_i - F_n) - \tfrac{1}{240}(\Delta''_i - \Delta''_n) + \tfrac{31}{60480}(\Delta^{\mathrm{iv}}_i - \Delta^{\mathrm{iv}}_n) - \cdot\;\cdot\;\cdot\}
\end{aligned}\right\} \tag{337}$$

$$\left.\begin{aligned}
'F_{\frac{1}{2}} &= \frac{H_0}{\omega} + \tfrac{1}{2}F_0 + \tfrac{1}{12}(\Delta'_0) - \tfrac{11}{720}(\Delta'''_0) + \tfrac{191}{60480}(\Delta^{\mathrm{v}}_0) - \cdot\;\cdot\;\cdot\;\cdot \\
\int\!\!\int_{t+n'\omega}^{t+n''\omega} F(T)\,dT^2 &= \omega^2\int\!\!\int_{n'}^{n''} F(t+n\omega)\,dn^2 \\
&= \omega^2\{(''F_{n''} - ''F_{n'}) + \tfrac{1}{12}(F_{n''} - F_{n'}) - \tfrac{1}{240}(\Delta''_{n''} - \Delta''_{n'}) + \tfrac{31}{60480}(\Delta^{\mathrm{iv}}_{n''} - \Delta^{\mathrm{iv}}_{n'}) - \cdot\;\cdot\}
\end{aligned}\right\} \tag{338}$$

In the preceding group the value of $''F_0$ is wholly arbitrary. We may, however, determine the quantity $''F_0$ such that the sum of the terms in (335) and (336) having the subscript *zero* will vanish : these formulae may therefore be written—

$$
\left.\begin{aligned}
'F_{\frac{1}{2}} &= \frac{H_0}{\omega} + \tfrac{1}{2}F_0 + \tfrac{1}{12}(\Delta'_0) - \tfrac{11}{720}(\Delta'''_0) + \tfrac{191}{60480}(\Delta^{v}_0) - \ldots \\
''F_0 &= -\tfrac{1}{12}F_0 + \tfrac{1}{240}\Delta''_0 - \tfrac{31}{60480}\Delta^{iv}_0 + \ldots \\
\int\int_t^{t+i\omega} F(T)\,dT^2 &= \omega^2 \int\int_0^i F(t+n\omega)\,dn^2 \\
&= \omega^2(''F_i + \tfrac{1}{12}F_i - \tfrac{1}{240}\Delta''_i + \tfrac{31}{60480}\Delta^{iv}_i - \ldots)
\end{aligned}\right\} \qquad (339)
$$

$$
\left.\begin{aligned}
'F_{\frac{1}{2}} &= \frac{H_0}{\omega} + \tfrac{1}{2}F_0 + \tfrac{1}{12}(\Delta'_0) - \tfrac{11}{720}(\Delta'''_0) + \tfrac{191}{60480}(\Delta^{v}_0) - \ldots \\
''F_0 &= -\tfrac{1}{12}F_0 + \tfrac{1}{240}\Delta''_0 - \tfrac{31}{60480}\Delta^{iv}_0 + \ldots \\
\int\int_t^{t+n\omega} F(T)\,dT^2 &= \omega^2 \int\int_0^n F(t+n\omega)\,dn^2 \\
&= \omega^2(''F_n + \tfrac{1}{12}F_n - \tfrac{1}{240}\Delta''_n + \tfrac{31}{60480}\Delta^{iv}_n - \ldots)
\end{aligned}\right\} \qquad (340)
$$

Let us now denote the second member of (332) by $\gamma(n)$; that is, let us put

$$\gamma(n) = ''F_n + \tfrac{1}{12}F_n - \tfrac{1}{240}\Delta''_n + \tfrac{31}{60480}\Delta^{iv}_n - \ldots \qquad (341)$$

Making $n = i + \frac{1}{2}$, this becomes

$$\gamma(i+\tfrac{1}{2}) = ''F_{i+\frac{1}{2}} + \tfrac{1}{12}F_{i+\frac{1}{2}} - \tfrac{1}{240}\Delta''_{i+\frac{1}{2}} + \tfrac{31}{60480}\Delta^{iv}_{i+\frac{1}{2}} - \ldots \qquad (342)$$

It will be observed from the foregoing schedule that $''F_{i+\frac{1}{2}}$, $F_{i+\frac{1}{2}}$, $\Delta''_{i+\frac{1}{2}}$, ... are not explicitly given, but must be derived from their respective series by interpolation *to halves*. For this purpose, let us put, in analogy with (269),

$$
\left.\begin{aligned}
(''F_{i+\frac{1}{2}}) &= \tfrac{1}{2}(''F_i + ''F_{i+1}) \qquad & (\Delta''_{i+\frac{1}{2}}) &= \tfrac{1}{2}(\Delta''_i + \Delta''_{i+1}) \\
(F_{i+\frac{1}{2}}) &= \tfrac{1}{2}(F_i + F_{i+1}) & & \ldots\ldots\ldots\ldots
\end{aligned}\right\} \qquad (343)
$$

then, after the manner of (270), we shall have

$$
\left.\begin{aligned}
''F_{i+\frac{1}{2}} &= (''F_{i+\frac{1}{2}}) - \tfrac{1}{8}(F_{i+\frac{1}{2}}) + \tfrac{3}{128}(\Delta''_{i+\frac{1}{2}}) - \tfrac{5}{1024}(\Delta^{iv}_{i+\frac{1}{2}}) + \ldots \\
F_{i+\frac{1}{2}} &= (F_{i+\frac{1}{2}}) - \tfrac{1}{8}(\Delta''_{i+\frac{1}{2}}) + \tfrac{3}{128}(\Delta^{iv}_{i+\frac{1}{2}}) - \ldots \\
\Delta''_{i+\frac{1}{2}} &= (\Delta''_{i+\frac{1}{2}}) - \tfrac{1}{8}(\Delta^{iv}_{i+\frac{1}{2}}) + \ldots \\
\Delta^{iv}_{i+\frac{1}{2}} &= (\Delta^{iv}_{i+\frac{1}{2}}) - \ldots \\
&\ldots\ldots\ldots\ldots\ldots\ldots
\end{aligned}\right\} \qquad (344)
$$

Upon substituting these expressions in the second member of (342), and reducing, we find

$$\gamma(i+\tfrac{1}{2}) = (''F_{i+\frac{1}{2}}) - \tfrac{1}{24}(F_{i+\frac{1}{2}}) + \tfrac{17}{1920}(\Delta''_{i+\frac{1}{2}}) - \tfrac{367}{193536}(\Delta^{iv}_{i+\frac{1}{2}}) + \ldots \qquad (345)$$

Again, by means of (332) and (341), we derive

$$\int\int_{n'}^{n''} F(t+n\omega)\,dn^2 = \gamma(n'') - \gamma(n') \tag{346}$$

Finally, denoting by $H_{-\frac{1}{2}}$ the value of $\int F(T)\,dT$ when $T = t - \frac{1}{2}\omega$, we shall have, by (328*a*),

$$\begin{aligned} H_{-\frac{1}{2}} &= \left[\int F(T)\,dT\right]_{T=t-\frac{1}{2}\omega} = \omega\left[\int F(t+n\omega)\,dn\right]_{n=-\frac{1}{2}} \\ &= \omega\left('F_{-\frac{1}{2}} + \tfrac{1}{24}\Delta'_{-\frac{1}{2}} - \tfrac{17}{5760}\Delta'''_{-\frac{1}{2}} + \tfrac{367}{967680}\Delta^{v}_{-\frac{1}{2}} - \ldots\ldots\right) \end{aligned}$$

which gives

$$'F_{-\frac{1}{2}} = \frac{H_{-\frac{1}{2}}}{\omega} - \tfrac{1}{24}\Delta'_{-\frac{1}{2}} + \tfrac{17}{5760}\Delta'''_{-\frac{1}{2}} - \tfrac{367}{967680}\Delta^{v}_{-\frac{1}{2}} + \ldots\ldots \tag{347}$$

By assigning various values to the limits n' and n'' in (346), and employing either (341) or (345) as required in each particular case; and finally, by using either (334) or (347) to determine the series $'F$, according as the assigned lower limit *is not* or *is* equal to $-\frac{1}{2}$, we derive the group of formulae given below :

$$\begin{aligned} 'F_{\frac{1}{2}} &= \frac{H_0}{\omega} + \tfrac{1}{2}F_0 + \tfrac{1}{12}(\Delta'_0) - \tfrac{11}{720}(\Delta'''_0) + \tfrac{191}{60480}(\Delta^{v}_0) - \ldots\ldots \\ ''F_0 &= -\tfrac{1}{12}F_0 + \tfrac{1}{240}\Delta''_0 - \tfrac{31}{60480}\Delta^{iv}_0 + \ldots\ldots \\ \int\int_t^{t+(i+\frac{1}{2})\omega} F(T)\,dT^2 &= \omega^2\int\int_0^{i+\frac{1}{2}} F(t+n\omega)\,dn^2 \\ &= \omega^2\left\{(''F_{i+\frac{1}{2}}) - \tfrac{1}{24}(F_{i+\frac{1}{2}}) + \tfrac{17}{1920}(\Delta''_{i+\frac{1}{2}}) - \tfrac{367}{193536}(\Delta^{iv}_{i+\frac{1}{2}}) + \ldots\ldots\right\} \end{aligned} \tag{348}$$

$$\begin{aligned} 'F_{\frac{1}{2}} &= \frac{H_0}{\omega} + \tfrac{1}{2}F_0 + \tfrac{1}{12}(\Delta'_0) - \tfrac{11}{720}(\Delta'''_0) + \tfrac{191}{60480}(\Delta^{v}_0) - \ldots\ldots \\ ''F_0 &= \textit{any convenient value; arbitrarily assigned.} \\ \int\int_{t+n\omega}^{t+(i+\frac{1}{2})\omega} F(T)\,dT^2 &= \omega^2\int\int_n^{i+\frac{1}{2}} F(t+n\omega)\,dn^2 \\ &= \omega^2\left\{(''F_{i+\frac{1}{2}}) - \tfrac{1}{24}(F_{i+\frac{1}{2}}) + \tfrac{17}{1920}(\Delta''_{i+\frac{1}{2}}) - \tfrac{367}{193536}(\Delta^{iv}_{i+\frac{1}{2}}) + \ldots\ldots \right. \\ &\qquad \left. - ''F_n - \tfrac{1}{12}F_n + \tfrac{1}{240}\Delta''_n - \tfrac{31}{60480}\Delta^{iv}_n + \ldots\ldots\right\} \end{aligned} \tag{349}$$

$$\begin{aligned} 'F_{-\frac{1}{2}} &= \frac{H_{-\frac{1}{2}}}{\omega} - \tfrac{1}{24}\Delta'_{-\frac{1}{2}} + \tfrac{17}{5760}\Delta'''_{-\frac{1}{2}} - \tfrac{367}{967680}\Delta^{v}_{-\frac{1}{2}} + \ldots\ldots \\ ''F_0 &= \tfrac{1}{2}\,'F_{-\frac{1}{2}} + \tfrac{1}{24}(F_{-\frac{1}{2}}) - \tfrac{17}{1920}(\Delta''_{-\frac{1}{2}}) + \tfrac{367}{193536}(\Delta^{iv}_{-\frac{1}{2}}) - \ldots\ldots \\ \int\int_{t-\frac{1}{2}\omega}^{t+i\omega} F(T)\,dT^2 &= \omega^2\int\int_{-\frac{1}{2}}^{i} F(t+n\omega)\,dn^2 \\ &= \omega^2\left(''F_i + \tfrac{1}{12}F_i - \tfrac{1}{240}\Delta''_i + \tfrac{31}{60480}\Delta^{iv}_i - \ldots\ldots\right) \end{aligned} \tag{350}$$

$$\left.\begin{aligned}
{}'F_{-\frac{1}{2}} &= \frac{H_{-\frac{1}{2}}}{\omega} - \tfrac{1}{24}\,\Delta'_{-\frac{1}{2}} + \tfrac{17}{5760}\,\Delta'''_{-\frac{1}{2}} - \tfrac{367}{967680}\,\Delta^{\mathrm{v}}_{-\frac{1}{2}} + \;.\;.\;.\;.\\
{}''F_0 &= \tfrac{1}{2}\,{}'F_{-\frac{1}{2}} + \tfrac{1}{24}\,(F_{-\frac{1}{2}}) - \tfrac{17}{1920}\,(\Delta''_{-\frac{1}{2}}) + \tfrac{367}{193536}\,(\Delta^{\mathrm{iv}}_{-\frac{1}{2}}) - \;.\;.\;.\;.\\
\iint_{t-\frac{1}{2}\omega}^{t+n\omega} F(T)\,dT^2 &= \omega^2 \iint_{-\frac{1}{2}}^{n} F(t+n\omega)\,dn^2\\
&= \omega^2({}''F_n + \tfrac{1}{12}\,F_n - \tfrac{1}{240}\,\Delta''_n + \tfrac{31}{60480}\,\Delta^{\mathrm{iv}}_n - \;.\;.\;.\;.\;)
\end{aligned}\right\} \quad (351)$$

$$\left.\begin{aligned}
{}'F_{-\frac{1}{2}} &= \frac{H_{-\frac{1}{2}}}{\omega} - \tfrac{1}{24}\,\Delta'_{-\frac{1}{2}} + \tfrac{17}{5760}\,\Delta'''_{-\frac{1}{2}} - \tfrac{367}{967680}\,\Delta^{\mathrm{v}}_{-\frac{1}{2}} + \;.\;.\;.\;.\\
{}''F_0 &= \textit{any convenient value; arbitrarily assigned.}\\
\iint_{t-\frac{1}{2}\omega}^{t+(i+\frac{1}{2})\omega} F(T)\,dT^2 &= \omega^2 \iint_{-\frac{1}{2}}^{i+\frac{1}{2}} F(t+n\omega)\,dn^2\\
&= \omega^2\Big[\{({}''F_{i+\frac{1}{2}}) - ({}''F_{-\frac{1}{2}})\} - \tfrac{1}{24}\,\{(F_{i+\frac{1}{2}}) - (F_{-\frac{1}{2}})\}\\
&\qquad + \tfrac{17}{1920}\,\{(\Delta''_{i+\frac{1}{2}}) - (\Delta''_{-\frac{1}{2}})\} - \tfrac{367}{193536}\,\{(\Delta^{\mathrm{iv}}_{i+\frac{1}{2}}) - (\Delta^{\mathrm{iv}}_{-\frac{1}{2}})\} + \;.\;.\;.\;.\Big]
\end{aligned}\right\} \quad (352)$$

The last formula may also be written in the following form:

$$\left.\begin{aligned}
{}'F_{-\frac{1}{2}} &= \frac{H_{-\frac{1}{2}}}{\omega} - \tfrac{1}{24}\,\Delta'_{-\frac{1}{2}} + \tfrac{17}{5760}\,\Delta'''_{-\frac{1}{2}} - \tfrac{367}{967680}\,\Delta^{\mathrm{v}}_{-\frac{1}{2}} + \;.\;.\;.\;.\\
{}''F_0 &= \tfrac{1}{2}\,{}'F_{-\frac{1}{2}} + \tfrac{1}{24}\,(F_{-\frac{1}{2}}) - \tfrac{17}{1920}\,(\Delta''_{-\frac{1}{2}}) + \tfrac{367}{193536}\,(\Delta^{\mathrm{iv}}_{-\frac{1}{2}}) - \;.\;.\;.\;.\\
\iint_{t-\frac{1}{2}\omega}^{t+(i+\frac{1}{2})\omega} F(T)\,dT^2 &= \omega^2 \iint_{-\frac{1}{2}}^{i+\frac{1}{2}} F(t+n\omega)\,dn^2\\
&= \omega^2\{({}''F_{i+\frac{1}{2}}) - \tfrac{1}{24}\,(F_{i+\frac{1}{2}}) + \tfrac{17}{1920}\,(\Delta''_{i+\frac{1}{2}}) - \tfrac{367}{193536}\,(\Delta^{\mathrm{iv}}_{i+\frac{1}{2}}) + \;.\;.\;.\;.\;\}
\end{aligned}\right\} \quad (353)$$

It may be well to again point out the fact that the functions and differences enclosed within parentheses denote the *means* of corresponding tabular quantities, as defined by (269) and (343). Further, that H_0 and $H_{-\frac{1}{2}}$ denote the values of the *first* integral of $F(T)$ when for T we substitute t and $t-\frac{1}{2}\omega$, respectively. Finally, we may add that if in any case H_p is given and H_q required, it is only necessary to compute

$$X = \int_{t+q\omega}^{t+p\omega} F(T)\,dT = H_p - H_q$$

and thence find

$$H_q = H_p - X \qquad (354)$$

In the process of double integration by mechanical quadrature it is sometimes convenient to tabulate, not the given function, but ω^2 times that quantity. By this means all differences are multiplied by ω^2, and thus the *final* multiplication by that factor is avoided. However, in order that the quantities $'F$ and $''F$ shall be multiplied by the same factor, it is evident that the independent term $\frac{H}{\omega}$ (which has the same fixed value whether we tabulate $F(T)$ or $\omega^2 F(T)$) must likewise be multiplied by ω^2: so that, proceeding by this method, it becomes necessary to take ωH in place of the term $\frac{H}{\omega}$ which occurs in all the preceding formulae. The computer is cautioned against neglecting this precept in case he tabulates $\omega^2 F(T)$ instead of the given function $F(T)$.

We close the chapter with several examples which illustrate the formulae given above.

EXAMPLE I.—Find the value of

$$Y = -\iint_{2.2}^{2.8} \frac{2T dT^2}{(1+T^2)^2}$$

on the supposition that the *first* integral vanishes for $T = 2.2$.

We tabulate the given function as below:

T	$''F$	$'F$	$F(T) \equiv \frac{-2T}{(1+T^2)^2}$	Δ'	Δ''	Δ'''	Δ^{iv}
2.0			−0.160000				
				+16499			
2.1			0.143501		−2009		
				14490		+263	
2.2	0.000000		0.129011		1746		−32
		−0.063375		12744		231	
2.3	−0.063375		0.116267		1515		32
		0.179642		11229		199	
2.4	0.243017		0.105038		1316		24
		0.284680		9913		175	
2.5	0.527697		0.095125		1141		28
		−0.379805		8772		147	
2.6	−0.907502		0.086353		994		−17
				7778		+130	
2.7			0.078575		− 864		
				+ 6914			
2.8			−0.071661				

Here we have

$$t = 2.2 \qquad \omega = 0.1 \qquad i = 4 \qquad H_0 = 0$$

whence, employing (335), we find

$$\begin{array}{llr} F_0 = -0.129011 & +\tfrac{1}{2} F_0 & = -0.0645055 \\ (\Delta'_0) = + \quad 13617 & +\tfrac{1}{12}(\Delta'_0) & = + \quad 1134.7 \\ (\Delta'''_0) = + \quad 247 & -\tfrac{11}{720}(\Delta'''_0) & = - \quad 3.8 \\ \hline & \therefore\ 'F_1 & = -0.063375 \end{array}$$

Assuming $''F_0 = 0$, we complete the table as shown above; thence, proceeding by (335), we obtain

$$\begin{array}{llll}
{}''F_4 = -0.907502 & {}''F_0 = 0.000000 & ({}''F_4 - {}''F_0) & = -0.907502 \\
F_4 = -0.086353 & F_0 = -0.129011 & +\frac{1}{12}(F_4 - F_0) & = + \quad 3554.8 \\
\Delta_4'' = - \quad 994 & \Delta_0'' = - \quad 1746 & -\frac{1}{240}(\Delta_4'' - \Delta_0'') & = - \quad 3.1 \\
\hline
 & & \Sigma & = -0.903950 \\
 & & \therefore\ Y & = -0.00903950
\end{array}$$

Verification: Integrating directly, we have

$$\int \frac{-2T dT}{(1+T^2)^2} = \frac{1}{1+T^2} + C$$

$$Y = \Big[\tan^{-1}T + CT\Big]_{2.2}^{2.6}$$

whence

$$0 = H_0 = \Big[(1+T^2)^{-1}\Big]_{T=2.2} + C$$

$$\therefore\ C = -0.17123288$$

Finally, using the relation

$$\tan^{-1}a - \tan^{-1}b = \tan^{-1}\left(\frac{a-b}{1+ab}\right)$$

the preceding expression for Y becomes

$$Y = \tan^{-1}\left(\frac{0.4}{6.72}\right) + 0.4C$$

which gives

$$Y = -0.00903949$$

EXAMPLE II.—From the table of the preceding example, compute

$$Y = -\iint_{2.23}^{2.55} \frac{2T dT^2}{(1+T^2)^2}$$

Here we employ (349), taking

$$t = 2.2 \qquad i = 3 \qquad H_0 = 0 \qquad n = (2.23-2.2) \div 0.1 = 0.30$$

Thus we find

$$\begin{array}{lll}
\ldots\ldots\ldots\ldots & ({}''F_{3\frac{1}{2}}) & = -0.717599.5 \\
(F_{3\frac{1}{2}}) = -0.090739 & -\frac{1}{24}(F_{3\frac{1}{2}}) & = + \quad 3780.8 \\
(\Delta_{3\frac{1}{2}}'') = - \quad 1068 & +\frac{17}{1920}(\Delta_{3\frac{1}{2}}'') & = - \quad 9.5 \\
\hline
 & \Sigma_1 & = -0.713828.2
\end{array}$$

Also

$$
\begin{array}{lll}
(n = 0.30) & & -\ {}''F_n = +0.006077.9 \\
F_n = -0.125016 & & -\ \tfrac{1}{12} F_n = +0.010418.0 \\
\Delta''_n = -\quad 1673 & & +\ \tfrac{1}{240}\Delta''_n = -\qquad 7.0 \\
\hline
 & & \Sigma_2 = +0.016488.9 \\
 & & \therefore\ \Sigma_1 + \Sigma_2 = -0.697339
\end{array}
$$

whence

$$Y = -0.00697339$$

Verifying this result as in the preceding example, we find

$$Y = \tan^{-1}\left(\frac{0.\dot{3}2}{6.6865}\right) + 0.32C = -0.00697338$$

Example III.—Let it be required to find

$$\mathbf{Y} = -\iint_{30°}^{50°} \frac{M \cos T dT^2}{\sin^2 T}$$

assuming that the *first integral* $= 2M$ when $T = 30°$; M being the modulus of the common system of logarithms.

Here we tabulate $F(T) \equiv -\omega^2 M \cos T \csc^2 T$ for $T = 20°, 24°, 28°, \ldots\ 60°$; thus avoiding the final multiplication by ω^2. Since $\omega = 4° = \pi \div 45$, we find

$$\log \omega^2 M = 7.325659 - 10$$

Our table is therefore as follows:

T	$''F$	$'F$	$F(T) \equiv -\omega^2 M \cos T \csc^2 T$	Δ'	Δ''	Δ'''	Δ^{iv}
20°			−0.017004				
				+5315			
24			0.011689		−2106		
				3209		+985	
28			0.008480		1121		−517
		+0.060553		2088		468	
32	+0.029974		0.006392		653		217
		0.054161		1435		251	
36	0.084135		0.004957		402		113
		0.049204		1033		138	
40	0.133339		0.003924		264		53
		0.045280		769		85	
44	0.178619		0.003155		179		30
		0.042125		590		55	
48	0.220744		0.002565		124		20
		+0.039560		466		35	
52	+0.260304		0.002099		89		− 12
				377		+ 23	
56			0.001722		− 66		
				+ 311			
60			−0.001411				

We proceed by formula (353), taking as our data

$$\begin{array}{llll} t = 32^\circ & & \omega = & 4^\circ = \pi \div 45 \\ i = 4 & & H_{-\frac{1}{2}} = & 2M = 0.868589 \end{array}$$

Whence, observing that we must now take $\omega H_{-\frac{1}{2}}$ instead of the term $H_{-\frac{1}{2}} \div \omega$ in (353), the computation of $'F_{-\frac{1}{2}}$ is as follows :

$$\begin{array}{rlr|rll} \log \omega H_{-\frac{1}{2}} = & 8.782752 & & \omega H_{-\frac{1}{2}} & = & +0.060639.0 \\ A'_{-\frac{1}{2}} = & +\ 2088 & & -\frac{1}{24} A'_{-\frac{1}{2}} & = & - \qquad 87.0 \\ A'''_{-\frac{1}{2}} = & +\ \ 468 & & +\frac{17}{5760} A'''_{-\frac{1}{2}} & = & + \qquad\ \ 1.4 \\ \hline & & & \therefore\ 'F_{-\frac{1}{2}} & = & +0.060553.4 \end{array}$$

And for $''F_0$ we find

$$\begin{array}{rl|rll} \cdot\ \cdot\ \cdot\ \cdot\ \cdot & \cdot\ \cdot\ \cdot\ \cdot\ \cdot & \frac{1}{2}\,'F_{-\frac{1}{2}} & = & +0.030276.7 \\ (F_{-\frac{1}{2}}) = & -0.007436 & +\frac{1}{24}(F_{-\frac{1}{2}}) & = & - \qquad 309.8 \\ (A''_{-\frac{1}{2}}) = & - \qquad 887 & -\frac{17}{1920}(A''_{-\frac{1}{2}}) & = & + \qquad\ \ 7.9 \\ (A^{\text{iv}}_{-\frac{1}{2}}) = & - \qquad 367 & +\frac{367}{193536}(A^{\text{iv}}_{-\frac{1}{2}}) & = & - \qquad\ \ 0.7 \\ \hline & & \therefore\ ''F_0 & = & +0.029974 \end{array}$$

Upon completing the table as shown above, and continuing the computation by (353), we obtain

$$\begin{array}{rl|rll} (i = & 4) & (''F_{4\frac{1}{2}}) & = & +0.240524.0 \\ (F_{4\frac{1}{2}}) = & -0.002332 & -\frac{1}{24}(F_{4\frac{1}{2}}) & = & + \qquad 97.2 \\ (A''_{4\frac{1}{2}}) = & - \qquad 106 & +\frac{17}{1920}(A''_{4\frac{1}{2}}) & = & - \qquad\ \ 0.9 \\ \hline & & \therefore\ Y & = & +0.240620\ . \end{array}$$

We easily verify this result analytically as follows :

$$\int \frac{-M \cos T dT}{\sin^2 T} = \frac{M}{\sin T} + C$$

$$\iint \frac{-M \cos T dT^2}{\sin^2 T} = M \log_e \tan \tfrac{1}{2} T + CT + C'$$

$$= \log_{10} \tan \tfrac{1}{2} T + CT + C'$$

$$\therefore\ Y = \Big[\log_{10} \tan \tfrac{1}{2} T + CT\Big]_{T = 30^\circ = \frac{\pi}{6}}^{T = 50^\circ = \frac{5}{18}\pi}$$

But

$$2M = H_{-\frac{1}{2}} = \frac{M}{\sin 30^\circ} + C = 2M + C$$

$$\therefore\ C = 0$$

$$\therefore\ Y = \log_{10} \tan\left(\frac{50^\circ}{2}\right) - \log_{10} \tan\left(\frac{30^\circ}{2}\right)$$

Now we find

$$\begin{array}{rcl} \log \tan 25^\circ & = & 9.668672.5 - 10 \\ \log \tan 15^\circ & = & 9.428052.5 - 10 \\ \hline \therefore\ Y & = & 0.240620 \end{array}$$

which agrees exactly with the former result.

Example IV.—From the table and data of Example III, compute the integral

$$Y = -\int\!\!\int_{30^\circ}^{45^\circ} \frac{M \cos T\, dT^2}{\sin^2 T}$$

Here we employ (351), taking $t = 32^\circ$ as before; we then have for the value of n at the upper limit,

$$n = (45^\circ - 32^\circ) \div 4^\circ = 3.25 = 3 + 0.25$$

We therefore obtain

$$\begin{array}{lcr@{\qquad}rcr} \cdot\ \cdot\ \cdot\ \cdot\ \cdot\ \cdot\ \cdot\ \cdot & & & {}^{\prime\prime}F_n & = & +0.189420.3 \\ F_n & = & -0.002993 & +\tfrac{1}{12} F_n & = & -\qquad 249.4 \\ \Delta_n'' & = & -\qquad 163 & -\tfrac{1}{240} \Delta_n'' & = & +\qquad 0.7 \\ \hline & & & \therefore\ Y & = & +0.189172 \end{array}$$

Verifying this result as in the last example, we find

$$Y = \log_{10} \tan 22^\circ\, 30' - \log_{10} \tan 15^\circ = +0.189172$$

Example V.—As a final exercise, combining both single and double integration, and illustrating, moreover, the use of formula (339) when several values are assigned in succession to the integer i, we shall conclude these examples with a complete and detailed solution of the following problem:

A particle P of unit mass is impelled along a straight line AB by a varying force whose expression is $20000\, T^{-3}$; where T is the time in seconds after a definite epoch, and the implied unit of length is one foot. It is required to find by quadratures the velocity, v, and the distance, $AP = x$, for the times

T = 102, 104, 106, 108 and 110 seconds, respectively;

assuming that $v_0 = 0.6$ feet per second and $x_0 = 8$ feet when $T_0 = 100$ seconds.

Since the mass of P is unity, we have, simply,

$$\frac{d^2x}{dT^2} = \frac{20000}{T^3}$$

whence by a single integration

$$v = \frac{dx}{dT} = \int_{T_0}^{T} \frac{20000dT}{T^3} + v_0 \qquad (\alpha)$$

and by double integration

$$x = \iint_{T_0}^{T} \frac{20000dT^2}{T^3} + x_0 \qquad (\beta)$$

We shall first compute the required values of x as given by equation (β), effecting the double integration by means of (339). The details of the computation are shown in the following table :

TABLE (A).

T	$F(T) \equiv 40000\,T^{-3}$	Δ'	Δ''	$'F$	$''F + \frac{1}{2}x_0 \equiv a$	$+\frac{1}{12}F \equiv b$	$\frac{1}{2}x = a + b$	x
96	0.04521							
		−271		+0.53730				
98	.04250		+21					
		250		.57980				
100	.04000		19		+3.99667	+0.00333	4.00000	8.0000
		231		.61980				
102	.03769		18		4.61647	314	4.61961	9.2392
		213		.65749				
104	.03556		15		5.27396	296	5.27692	10.5538
		198		.69305				
106	.03358		15		5.96701	280	5.96981	11.9396
		183		.72663				
108	.03175		13		6.69364	265	6.69629	13.3926
		170		.75838				
110	.03005		12		+7.45202	+0.00250	7.45452	14.9090
		158		.78843				
112	.02847		+11					
		−147		+0.81690				
114	0.02700							

Since we shall afterwards use this same table in finding v by single integration, it is here convenient to tabulate ω times the given function : thus avoiding the final multiplication by ω in computing v, and reducing the corresponding factor in the case of x from ω^2 to ω. Accordingly, we tabulate under $F(T)$ the function

$$F(T) \equiv 20000\omega T^{-3} = 40000\,T^{-3}$$

Assume $t = 100$, and proceed by (339). To determine $'F_{\frac{1}{2}}$, it must be observed that since $F(T)$, Δ', Δ'', ... already contain the factor ω, it is here necessary to multiply the independent term $\frac{H_0}{\omega}$

by the same factor: so that, writing $v_0 (= H_0)$ for $\frac{H_0}{\omega}$ in the first equation of (339), and omitting insensible terms, we have

$$'F_1 = v_0 + \tfrac{1}{2} F_0 + \tfrac{1}{12} (\Delta'_0) \qquad (\gamma)$$

Hence, substituting $v_0 = 0.6$, $F_0 = 0.04000$, $(\Delta'_0) = \frac{1}{2}(\Delta'_{-1} + \Delta'_1) = -0.00240$, we find $'F_{\frac{1}{2}} = +0.61980$, and thus complete the series $'F$ as given above.

The second equation of (339) gives simply, $''F_0 = -\frac{1}{12} F_0$, the term in Δ'' being insensible. But since, by equation (β), we should afterwards have to add the constant x_0 to each computed value of the double integral taken from T_0 to T, it is expedient to tabulate in place of $''F_0$ the quantity

$$''F_0 + \frac{x_0}{\omega} = ''F_0 + \tfrac{1}{2} x_0 = -\tfrac{1}{12} F_0 + 4.0 = 4.0 - 0.00333 = +3.99667$$

and thence complete the series as given under $''F + \frac{1}{2} x_0 \equiv a$. The reason for this procedure is easily made apparent: for the final equation of (339) gives (since ω^2 must now be replaced by ω)

$$\iint_{T_0}^{T} \frac{20000 dT^2}{T^8} = \omega (''F_i + \tfrac{1}{12} F_i)$$

and substituting this expression in equation (β), we obtain

$$x = \omega (''F_i + \tfrac{1}{12} F_i) + x_0 = \omega \left(''F_i + \frac{x_0}{\omega} + \tfrac{1}{12} F_i\right) \qquad (\delta)$$

Therefore, upon forming the column $+\frac{1}{12} F \equiv b$, as given above, we have from (δ)

$$\tfrac{1}{2} x = ''F_i + \tfrac{1}{2} x_0 + \tfrac{1}{12} F_i = a + b$$

whence the required values of x are derived and tabulated in the final column of Table (A).

For the computation of the velocity v we employ formula (282), the first equation of which gives

$$'F_{-\frac{1}{2}} = -\tfrac{1}{2} F_0 + \tfrac{1}{12}(\Delta'_0)$$

or, by adding F_0 to both members,

$$'F_{\frac{1}{2}} = +\tfrac{1}{2} F_0 + \tfrac{1}{12}(\Delta'_0)$$

But we shall avoid subsequent additions of the constant v_0, required by equation (a), if we increase this value of $'F_{\frac{1}{2}}$ by the term $v_0 = 0.6$; that is, if we take

$$'F_{\frac{1}{2}} = v_0 + \tfrac{1}{2} F_0 + \tfrac{1}{12}(\Delta'_0)$$

which is the same as the expression (γ), used for determining the series $'F$ in Table (A). The latter series is therefore to be employed in finding v, the computation of which is as follows:

TABLE (B).

T	$('F)$	(Δ')	$-\frac{1}{12}(\Delta')$	$v=('F)-\frac{1}{12}(\Delta')$
96	+0.51470	. .	+24	+0.51494
98	.55855	−260	22	.55877
100	.59980	240	20	.60000
102	.63865	222	18	.63883
104	.67527	205	17	.67544
106	.70984	190	16	.71000
108	.74251	176	15	.74266
110	.77341	164	14	.77355
112	.80267	−152	13	.80280
114	+0.83040	. .	+12	+0.83052

Recalling the fact that functions and differences in parentheses are *means* taken according to (269), the method of forming the second, third and fourth columns of this table from the quantities of Table (A) is obvious. Now, since the factor ω has been previously introduced, the second equation of (282) gives

$$v = ('F_i) - \tfrac{1}{12}(\Delta'_i)$$

from which expression the required values of v are computed and tabulated in the final column of Table (B).

This completes the solution of the problem. An interesting check is derived, however, by observing that equation (a) gives

$$x = \int_{T_0}^{T} v\,dT + x_0 \qquad (\epsilon)$$

whence x may be obtained from the series v by single integration. For this purpose we make $f(T) \equiv \omega v = 2v$, and thus form the table below :

TABLE (C).

T	$f(T) \equiv 2v$	δ'	δ''	$'f + x_0$	$('f) + x_0 \equiv c$	(δ')	$-\frac{1}{12}(\delta') \equiv d$	$x = c + d$
96	1.0299							
		+876						
98	1.1175		−51					
		825		+ 7.4067				
100	1.2000		48		8.0067	+801	−67	8.0000
		777		8.6067				
102	1.2777		45		9.2455	754	63	9.2392
		732		9.8844				
104	1.3509		41		10.5598	711	59	10.5539
		691		11.2353				
106	1.4200		38		11.9453	672	56	11.9397
		653		12.6553				
108	1.4853		35		13.3979	636	53	13.3926
		618		14.1406				
110	1.5471		33		14.9141	+602	−50	14.9091
		585		+15.6877				
112	1.6056		−31					
		+554						
114	1.6610							

Here again we take $t = 100$, and employ (282), which gives

$$'f_{-\frac{1}{2}} = -\tfrac{1}{2} f_0 + \tfrac{1}{12}(\delta'_0) = -0.6000 + 0.0067 = -0.5933$$

Increasing this value by $x_0 = 8.0$, to provide for the constant x_0 in equation (ϵ), we get $+7.4067$, which number is written under $'f + x_0$, on the line $t - \frac{1}{2}\omega$. Completing this column by successive additions of the functions f, we next form the series of *mean* values tabulated under $('f) + x_0 \equiv c$. The columns (δ') and $-\frac{1}{12}(\delta') \equiv d$ are then computed, and finally the column $x = c + d$. These values of x agree substantially with those given in Table (A).

From the given analytical expression for the force, together with the initial conditions of the problem, we easily find

$$v = 1.6 - 10000\,T^{-2} \quad , \quad x = 1.6\,T + 10000\,T^{-1} - 252$$

whence, making $T = 110$, we obtain

$$v = 0.77355 \quad \text{and} \quad x = 14.9091$$

which further verify the results derived by quadratures.

79. It is worth while to inquire what change takes place in the value of the double integral

$$Y = \int\int_{T'}^{T''} F(T)\, dT^2$$

when, in a particular problem, the quantity H is changed from an assigned value H' to a new value H''. This is easily answered. For, if we change H' to H'', the value of the first integral — corresponding to *any* particular value of T — is thereby increased by the quantity $H''—H'$; or, what amounts to the same thing, the *constant* of the first integration, M in (286*a*), is thus increased by $H''—H'$. Therefore, by (287), it is evident that Y is increased by the quantity $(H''—H')\,(T''—T')$.

EXAMPLES.

1. Given the semi-major axis of an ellipse, $a = 1$, and the semi-minor axis, $b = 0.8$, to find the length of the elliptic quadrant.

Ans. 1.41808.

[NOTE: — Take the eccentric angle E as independent variable, and hence find

$$s = \int_0^{\frac{\pi}{2}} \sqrt{1 - e^2 \cos^2 E}\, dE$$

where e is the eccentricity, and s the required length.]

2. Given the equation of a cardioid, $r = 1 + \cos\theta$: to find, by mechanical quadrature, the length of that part of the curve comprised between the initial line and a line through the pole at right-angles to the initial line.

Ans. 2.82843.

3. The equation of a curve being $y = x^2\sqrt{2 - \sin x}$, find the area included between the curve, the axis of x, and the two ordinates, $x = \frac{\pi}{3}$ and $x = \frac{2}{7}\pi$.

Ans. 0.180518.

4. Compute the value of

$$Y = \iint_0^{\frac{\pi}{6}} \frac{dT^2}{\sqrt{1 - 0.82 \sin^2 T}}$$

assuming that the *first* integral vanishes at the lower limit.

Ans. 0.139727.

5. Given a curve in a vertical plane whose points satisfy the relation

$$\frac{d^2y}{dx^2} = \frac{4x^2 - 3}{5 + \sqrt{x}}$$

— the axis of y being vertical. Find the difference of level between two points whose abscissae are 1.000 and 1.473, respectively; assuming the direction of the curve to be horizontal at the first point.

Ans. 0.044228.

6. By what amount would the preceding result be changed by supposing the tangent to the curve at the first point to be inclined 45° to the horizontal?

[NOTE: — This question should be answered mentally.]

CHAPTER V.

MISCELLANEOUS PROBLEMS AND APPLICATIONS.

80. The present short chapter will be devoted to the solution of a number of problems and examples involving certain principles and precepts hitherto established.

81. PROBLEM I.—*To find* $S \equiv 1^k + 2^k + 3^k + \ldots + r^k$, *where* k *and* r *are integers.*

The method of solution is best illustrated by assigning a particular value to k. Thus, let it be required to find

$$S \equiv 1^4 + 2^4 + 3^4 + \ldots + r^4$$

We tabulate below and difference the values of T^4 which correspond to $T = 1, 2, 3, 4, 5$ and 6. Thus we find :

T	$'F$	$F(T) \equiv T^4$	Δ'	Δ''	Δ'''	Δ^{iv}	Δ^{v}
	$'F_0$						
1	$'F_1$	1					
			15				
2	$'F_2$	16		50			
			65		60		
3		81		110		24	
	.		175		84		0
4		256		194		24	
	.		369		108		.
5		625		302		.	
			671		.		.
6		1296		.		.	
			.		.		.
	.			.		.	
.	.		.		.		.
				.		.	
.	.	.			.		
$r-1$	$'F_{r-1}$	$(r-1)^4$					
			.				
r	$'F_r$	r^4					

Now, by Theorem V, the 4th differences of $F(T)$ are constant, and hence the 5th and higher differences all *vanish*. Whence, if we consider the auxiliary series $'F$—defined as in Chapter IV—we shall have, by the fundamental formula (73),

$$'F_r = 'F_0 + r + \frac{r(r-1)}{\underline{|2}}(15) + \frac{r(r-1)(r-2)}{\underline{|3}}(50)$$

$$+ \frac{r(r-1)(r-2)(r-3)}{\underline{|4}}(60) + \frac{r(r-1)\ldots\ldots(r-4)}{\underline{|5}}(24)$$

$$= 'F_0 + \frac{r}{30}(r+1)(2r+1)(3r^2+3r-1)$$

Therefore, by Theorem I, we have

$$S = 'F_r - 'F_0 = \frac{r}{30}(r+1)(2r+1)(3r^2+3r-1) \tag{355}$$

which is the required expression for the sum of the fourth powers of the first r integers.

82. Problem II.— *Given a series of functions, F_{-3}, F_{-2}, F_{-1}, F_0, F_1, F_2, ... , and an assigned intermediate value, F_n: To find the corresponding interval n.*

First Solution: The simplest method is to determine by inspection an *approximate* value of n, and then find by direct interpolation the values of the function corresponding to three or four closely equidistant values of n that shall embrace the required interval. The latter is then readily found by a simple interpolation.

Example.—From the following ephemeris find the time when the logarithm of *Mercury's* distance from the Earth = 9.7968280: that is, given F_n = 9.7968280, to find n. The tabular quantities are here given for every second Greenwich mean noon.

Date 1898	Log. Dist. of ☿ from ⊕	Δ'	Δ''	Δ'''	Δ^{iv}	Δ^{v}
May 8	9.7560706					
		+ 91669				
10	9.7652375		+24839			
		116508		−4748		
12	9.7768883		20091		+382	
		136599		4366		+ 75
14	9.7905482		15725		457	
		152324		3909		+135
16	9.8057806		11816		+592	
		164140		−3317		
18	9.8221946		+ 8499			
		+172639				
20	9.8394585					

We observe that the given logarithm falls somewhere between the tabular values for May 14 and 16, and soon find that the interval (from the former date) is somewhat greater than 0.4. Hence we take $F_0 = 9.7905482$, and interpolate — by BESSEL's Formula — the functions corresponding to $n = 0.38$, 0.41, and 0.44. Thus, computing and differencing these values, we find

n	F_n	Δ'	Δ''
0.38	9.7961736		
		+4531	
0.41	9.7966267		+12
		+4543	
0.44	9.7970810		

Whence, if we denote by n' the interval at which the required function lies beyond the middle function in this new series, we shall have, by neglecting the small second difference,

$$n' = 2013 \div 4543 = 0.44, \text{ nearly.}$$

But if great accuracy is required, we may easily take account of the second difference by the method of the *corrected first difference* (§ 44). Thus, in the last table, we find that the corrected first difference which corresponds to $n' = 0.44$ is 4540; hence we have

$$n' = 2013 \div 4540 = 0.4434$$
$$\therefore n = 0.41 + 0.4434 \times 0.03 = 0.423302$$

The required time is, therefore,

$$T = \text{May } 14^d + 0.423302 \times 48^h = \text{May } 14^d\ 20^h\ 19^m\ 6^s.6$$

83. *Second Solution of Problem II.* — Given F_n, to find the value of n.

Let m denote an approximate value of n, true to the *nearest tenth* of a unit, and put

$$n = m + z \tag{356}$$

Then we have

$$F_n = F_{m+z} = F[t+(m+z)\omega] = F[(t+m\omega)+z\omega]$$
$$= F(t+m\omega) + z\omega F'(t+m\omega) + \frac{z^2\omega^2}{\underline{|2}} F''(t+m\omega) + \ldots .$$

Since we have supposed z not to exceed 0.05, it is permissible to neglect z^3, z^4, ... in the last expression, which becomes, therefore,

$$F_n = F_m + z\omega F'_m + \tfrac{1}{2} z^2\omega^2 F''_m \tag{357}$$

To find z from this equation, we first neglect the small term in z^2, and thus obtain an approximate value which we shall call x. In this manner we find

$$x = \frac{F_n - F_m}{\omega F'_m} \tag{358}$$

This approximate value of z will now suffice for substitution in the last term of (357). Accordingly, we obtain

$$z = x - \tfrac{1}{2} x^2 \left(\frac{\omega^2 F''_m}{\omega F'_m} \right) \tag{359}$$

whence, putting

$$y = \tfrac{1}{2} x^2 \left(\frac{\omega^2 F''_m}{\omega F'_m} \right) \tag{360}$$

we have

$$z = x - y$$

and equation (356) becomes

$$n = m + x - y \tag{361}$$

Finally, to express F_m, $\omega F'_m$, and $\omega^2 F''_m$ in terms of the differences of the given series F, it will be expedient to employ STIRLING'S Formula of interpolation, together with the expressions for F'_m and F''_m as developed in § 61. The above solution may then be expressed as follows :

$$\left.\begin{array}{lll}
\text{Determine} & m = & \text{an approximate value of } n\text{, true to the } \textit{nearest tenth} \\
 & & \quad \text{of a unit.} \\
\text{Thence find} & F_m = & F_0 + ma + Bb_0 + Cc + Dd_0 + \; . \; . \; . \; . \\
 & D_1 \equiv \omega F'_m = & a + mb_0 + C'c + D'd_0 + \; . \; . \; . \; . \\
 & D_2 \equiv \omega^2 F''_m = & b_0 + mc + \; . \; . \; . \; . \\
 & K = & \dfrac{D_2}{D_1} \\
 & x = & \dfrac{F_n - F_m}{D_1} \\
 & y = & \tfrac{1}{2} x^2 K \\
\text{and} & n = & m + x - y
\end{array}\right\} \tag{362}$$

Here the differences are to be taken according to the schedule on page 62 ; the coefficients B, C, D, ... being taken from Table II, and C', D', ... from Table V. Finally, Table VII gives the value of y for top argument K and side argument x ; observing that y has the same sign as K.

EXAMPLE. — Same as in §82.

Here we find $m = 0.40$; and hence take from the given table, and from Tables II and V, the quantities

$m = 0.40$	$a = +144461.5$	. . . .
$B = +0.080$	$b_0 = +\ 15725$	. . . .
$C = -0.056$	$c = -\ 4137.5$	$C' = -0.08667$
$D = -0.0056$	$d_0 = +\ 457$	$D' = -0.02267$
$E = +0.01075$	$e = +\ 105$	$E' = +0.01440$

The computation of F_m, D_1 and D_2 by (362) is therefore as follows:

$F_0 = 9.7905482$	. . . .	. . . .
$ma = +\ 57784.6$	$a = +144461.5$	. . . .
$Bb_0 = +\ 1258.0$	$mb_0 = +\ 6290.0$	$b_0 = +15725$
$Cc = +\ 231.7$	$C'c = +\ 358.6$	$mc = -\ 1655$
$Dd_0 = -\ 2.6$	$D'd_0 = -\ 10.4$	. . . .
$Ee = +\ 1.1$	$E'e = +\ 1.5$	. . . .
$\therefore F_m = 9.7964755$	$\therefore D_1 = +151101$	$\therefore D_2 = +14070$
$F_n = 9.7968280$		$\therefore F_n - F_m = +\ 3525$

Whence

$$K = D_2 \div D_1 = +14070 \div 151101 = +0.0931$$
$$x = (F_n - F_m) \div D_1 = +3525 \div 151101 = +0.023329$$

and we finally obtain

$$\begin{aligned} m &= 0.400000 \\ x &= +0.023329 \\ \text{(Table VII)}\ -y &= -\ \ \ 26 \\ \hline \therefore n &= 0.423303 \end{aligned}$$

which agrees within one unit with the former result.

84. PROBLEM III.— *To solve any numerical equation whatever involving but one unknown quantity.*

The given equation, whether simple or complex, algebraic or transcendental, may be written in the form

$$F(T) = 0$$

The problem therefore reduces to the question of finding n when F_n is known and equal to zero — which is the same as Problem II.

EXAMPLE. — Solve the transcendental equation

$$T - 20° \sin T = 45°$$

where T is expressed in degrees of arc.

This equation may be written

$$F(T) \equiv T - 20° \sin T - 45° = 0$$

which by trial we find to be satisfied by a value of T not far from 63°; hence we tabulate $F(T)$ for $T = 62°$, 63°, and 64°, as follows:

T	$F(T)$	Δ'	Δ''
62°	−0.6590		
		+8389	
63	+0.1799		+53
		+8442	
64	+1.0241		

Here we have given $F_n = 0$, to find n. Whence, employing the *corrected first difference* (§ 45), we find

$$T = 63° - \frac{1799}{8410} \times 1° = 62°.7861$$

85. PROBLEM IV. — *Given a series of numerical functions embracing a maximum or minimum value: To find the value of the argument which corresponds to the maximum or minimum function.*

Find by inspection the tabular function which falls *nearest* the required maximum or minimum value. Call this tabular function F_0. Then, from the schedule

T	$F(T)$	Δ'	Δ''	Δ'''	Δ^{iv}
$t - \omega$	F_{-1}		b'		d'
		a'		c'	
t	F_0		b_0		d_0
		a_1		c_1	
$t + \omega$	F_1		b_1		d_1

we have, by the first of equations (182),

$$F'(T) = F'(t+n\omega)$$
$$= \frac{1}{\omega}\left[(a - \tfrac{1}{6}c + \ldots) + n(b_0 - \tfrac{1}{12}d_0 + \ldots) + \tfrac{1}{2}n^2(c - \ldots) + \tfrac{1}{6}n^3(d_0 - \ldots) + \ldots\right]$$

Therefore, since the condition of maximum or minimum requires that $F'(T) = 0$, we have, by neglecting 5th differences,

$$(a-\tfrac{1}{6}c) + (b_0-\tfrac{1}{12}d_0)n + \tfrac{1}{2}cn^2 + \tfrac{1}{6}d_0n^3 = 0 \qquad (363)$$

which determines the value of n, and hence, also, the value of T, at the point of maximum or minimum of $F(T)$. This equation may be readily solved by successive approximations, by first neglecting the terms containing n^2 and n^3, and afterwards substituting therein the approximate value of n thus found, and so on; or, we may consider the solution of (363) from the standpoint of Problem III, — which may be regarded as the more direct of the two methods.

EXAMPLE. — The following ephemeris gives the log radius vector of *Mars* with respect to the Sun ($\log r$). Find the time of perihelion passage of the planet.

Date 1898	Log r	Δ'	Δ''	Δ'''	Δ^{iv}
April 6	0.1416628				
		−7325			
14	0.1409303		+2844		
		4481		+47	
22	0.1404822		2891		−27
		−1590		+20	
30	0.1403232		2911		33
		+1321		−13	
May 8	0.1404553		2898		−36
		4219		−49	
16	0.1408772		+2849		
		+7068			
24	0.1415840				

Here we are required to find the instant when $\log r$ is a minimum. Since it is evident that this condition occurs only a few hours from April 30, we take $F_0 = 0.1403232$. Whence, from the above table, we find

$$\begin{aligned} a &= -134.5 & a-\tfrac{1}{6}c &= -135 \\ b_0 &= +2911 & b_0-\tfrac{1}{12}d_0 &= +2914 \\ c &= +\ \ 3.5 & \tfrac{1}{2}c &= +\ \ 2 \\ d_0 &= -\ \ 33 & \tfrac{1}{6}d_0 &= -\ \ 6 \end{aligned}$$

and therefore, by (363),

$$-135 + 2914n + 2n^2 - 6n^3 = 0$$

or

$$2914n = 135 - 2n^2 + 6n^3$$

Neglecting the last two terms of this equation, we have, for an approximate value of n,

$$n = 135 \div 2914 = 0.046, \quad \text{nearly};$$

and since for this value of n the small terms sensibly vanish, we obtain as our final value

$$n = 135 \div 2914 = 0.04633$$

The date of perihelion passage is, therefore,

$$T = \text{April } 30^{d} + 0.04633 \times 8 \times 24^{h} = \text{April } 30^{d}\ 8^{h}.895$$

86. PROBLEM V.—*Given a series of numerical values* (F_{-3}, F_{-2}, F_{-1}, F_0, F_1, F_2, ...) *of any function* $F(T)$ *which is analytically unknown: To find an approximate algebraic expression for* $F(T)$ *in terms of the variable argument.*

Let us put

$$\tau = T - t \tag{364}$$

and TAYLOR'S Theorem gives

$$F(T) = F(t+\tau) = F(t) + \tau F'(t) + \frac{\tau^2}{\underline{|2}} F''(t) + \frac{\tau^3}{\underline{|3}} F'''(t) + \ldots . \tag{365}$$

Upon substituting in (365) the expressions for $F'(t)$, $F''(t)$, $F'''(t)$, ... , as given by (175), we obtain

$$\begin{aligned} F(T) = F(t) &+ \frac{1}{\omega}\left(a - \tfrac{1}{6}c + \tfrac{1}{30}e - \ldots .\right)\tau + \frac{1}{\omega^2\underline{|2}}\left(b_0 - \tfrac{1}{12}d_0 + \ldots .\right)\tau^2 \\ &+ \frac{1}{\omega^3\underline{|3}}\left(c - \tfrac{1}{4}e + \ldots .\right)\tau^3 + \frac{1}{\omega^4\underline{|4}}\left(d_0 - \ldots .\right)\tau^4 + \frac{1}{\omega^5\underline{|5}}\left(e - \ldots .\right)\tau^5 + \ldots . \end{aligned} \tag{366}$$

which expresses $F(T)$ as a rational integral function of τ, with known numerical coefficients; τ being the value of the variable argument counted from the fixed epoch t, as defined by (364).

EXAMPLE.—From NEWCOMB'S *Astronomical Constants* we take the following table of the mean obliquity of the ecliptic (ϵ) for every fifth century:

Year	Obliquity	Δ'	Δ''	Δ'''
	° ′ ″	′ ″	″	″
0	23 41 43.78			
		−3 45.81		
500	37 57.97		−4.09	
		3 49.90		+1.35
1000	34 8.07		2.74	
		3 52.64		1.36
1500	30 15.43		1.38	
		3 54.02		+1.36
2000	26 21.41		−0.02	
		−3 54.04		
2500	23 22 27.37			

Let it be required to express ϵ in terms of τ, the latter being counted from the year 1000 in terms of a century as the unit.

Since we adopt one century as the unit of time, it is necessary to express ω in the same unit ; therefore we have

$$\omega = 5 \qquad t = 1000^y \qquad F(t) = 23°\ 34'\ 8''.07$$

$$a = -3'\ 51''.27 = -231''.27 \qquad b_0 = -2''.74 \qquad c = +1''.355$$

$$a - \tfrac{1}{6}c = -231''.496 \qquad \omega^2 \underline{|2} = 50 \qquad \omega^3 \underline{|3} = 750$$

Whence, by (366), we obtain

$$\text{Coefficient of } \tau = -231''.496 \div 5 = -46''.299$$
$$\text{" " } \tau^2 = -\ 2.74 \div 50 = -\ 0.0548$$
$$\text{" " } \tau^3 = +\ 1.355 \div 750 = +\ 0.00181$$

Accordingly, the required expression for the obliquity is—

$$\epsilon = 23°\ 34'\ 8''.07 - 46''.299\,\tau - 0''.0548\,\tau^2 + 0''.00181\,\tau^3$$

Verification : Putting $\tau = 10$ in this formula, we should get the obliquity for 2000. Now we find

$$\text{(For 2000)} \quad \epsilon = 23°\ 34'\ 8''.07 - 462''.99 - 5''.48 + 1''.81 = 23°\ 26'\ 21''.41$$

which agrees exactly with the tabular value above.

It will be observed that the solution given by (366) restricts the *epoch*, or origin from which τ is counted, to some *tabular* value of the argument, as t. Should the assigned epoch be some *intermediate* value of T, say T_1, it will only be necessary to write

$$\tau_1 = T - T_1$$

and we have

$$F(T) = F(T_1 + \tau_1) = F(T_1) + \tau_1 F'(T_1) + \frac{\tau_1^2}{\underline{|2}} F''(T_1) + \ldots .$$

Therefore, if we put

we shall have
$$\left.\begin{aligned} T_1 &= t + m\omega \\ F(T) &= F_m + \tau_1 F'_m + \frac{\tau_1^2}{\lfloor 2} F''_m + \frac{\tau_1^3}{\lfloor 3} F'''_m + \ldots \end{aligned}\right\} \qquad (366a)$$

where $\tau_1 (= T - T_1)$ is the value of the variable argument counted from the assigned epoch T_1. Accordingly, if we compute by the usual methods the values of $F_m, F'_m, F''_m, F'''_m, \ldots,$ and substitute these in (366*a*), we shall obtain the expression required.

As an example, let us express the obliquity (ϵ) as a function of the time (τ_1) counted from the epoch 1600.0 in terms of a century as the unit.

Reverting to the above table, we take

$$t = 1500^y \qquad T_1 = 1600^y \qquad m = 0.20$$

Whence we find

$$F_m = 23^\circ\ 29'\ 28''.69 \qquad F'_m = -46''.761 \qquad F''_m = -0''.0443 \qquad F'''_m = +0''.01088$$

Substituting these values in the formula (366*a*), we obtain the required expression, namely,

$$\epsilon = 23^\circ\ 29'\ 28''.69\ -46''.761\,\tau_1\ -0''.0222\,\tau_1^2\ +0''.00181\,\tau_1^3$$

87. Geometrical Problem.—A circular well four feet in diameter is centrally intersected by a horizontal cylindrical shaft whose diameter is one foot. Find the volume of the portion of the shaft within the well.

Solution: Consider a vertical section or lamina of the shaft parallel to its axis, at a horizontal distance x from the latter, and having the differential thickness dx. Then, if we denote the radii of well and shaft by R and r, respectively, we shall have for the length of this rectangular section

$$l = 2\sqrt{R^2 - x^2}$$

and for its breadth, or height,

$$h = 2\sqrt{r^2 - x^2}$$

Therefore, the volume of the differential section is —

$$dV = lhdx = 4\sqrt{(R^2-x^2)(r^2-x^2)}\,dx$$

whence

$$V = 8\int_0^r \sqrt{(R^2-x^2)(r^2-x^2)}\,dx$$

Upon substituting the given values of R and r in this formula, it becomes

$$V = 8\int_0^{\frac{1}{2}} \sqrt{(4-x^2)(\tfrac{1}{4}-x^2)}\,dx$$

This expression belongs to the class of functions known as elliptic integrals, and therefore cannot be integrated directly. Accordingly, we proceed to evaluate V by mechanical quadrature. For this purpose it will be convenient to put

$$x = \tfrac{1}{2}\sin\theta$$

whence

$$dx = \tfrac{1}{2}\cos\theta d\theta$$

and the preceding expression for V becomes

$$V = \int_0^{\frac{\pi}{2}} \cos^2\theta\sqrt{16-\sin^2\theta}\,d\theta \qquad (367)$$

We now tabulate $F(\theta) \equiv \omega\cos^2\theta\sqrt{16-\sin^2\theta}$ (where $\omega = 10° = \pi \div 18$) as follows :

θ	$'F$	$F(\theta)$	Δ'	Δ''	Δ'''	Δ^{iv}
−15°		0.6500		−371		
			+ 427		− 56	
− 5		0.6927		427		+56
	0.0000		0		0	
+ 5		0.6927		427		56
	0.6927		− 427		+ 56	
15		0.6500		371		47
	1.3427		798		103	
25		0.5702		268		31
	1.9129		1066		134	
35		0.4636		−134		+12
	2.3765		1200		146	
45		0.3436		+ 12		− 4
	2.7201		1188		142	
55		0.2248		154		22
	2.9449		1034		120	
65		0.1214		274		37
	3.0663		760		83	
75		0.0454		357		37
	3.1117		− 403		+ 46	
85		0.0051		403		46
	3.1168		0		0	
95		0.0051		403		−46
			+ 403		− 46	
+105		0.0454		+357		

Accordingly, we take

$$t = 5^\circ \qquad i = 8 \qquad t + i\omega = 85^\circ$$

and proceed by formula (259): thus, observing that $\Delta'_{-\frac{1}{2}}, \Delta'''_{-\frac{1}{2}}, \ldots$ and $\Delta'_{i+\frac{1}{2}}, \Delta'''_{i+\frac{1}{2}}, \ldots$ are *all zero*, and remembering that the factor ω has already been introduced, we find

$$'F_{-\frac{1}{2}} = 0$$

and

$$V = {'F}_{i+\frac{1}{2}} = 3.1168 \text{ cubic feet}$$

88. Various other problems and applications of a similar nature might be added; indeed, Astronomy itself presents a large variety of such. But the leading principles of our subject have already been developed, explained, and exemplified. We therefore feel confident in leaving the student who has thoroughly mastered these principles, believing him fully capable of solving any further questions or problems that may arise in his practice.

EXAMPLES.

1. Derive the expression for the sum of the cubes of the first r integers.

Ans. $\frac{1}{4}r^2(r+1)^2$.

2. Find from the following ephemeris the instant when *Autumn commences*; that is, the instant when the Sun's right-ascension (α) equals twelve hours.

Date 1898	Sun's R.A. α	Date 1898	Sun's R.A. α
	h m s		h m s
Sept. 13	11 25 47.56	Sept. 25	12 8 54.44
16	11 36 33.99	28	12 19 43.35
19	11 47 20.29	Oct. 1	12 30 34.30
22	11 58 6.94	4	12 41 27.92

Ans. Sept. 22^d 12^h $34^m.8$.

3. From the ephemeris of the moon's latitude given below, determine the instant of greatest latitude north.

Date 1898	Moon's Latitude	Date 1898	Moon's Latitude
d	° ′ ″	d	° ′ ″
July 9.0	+5 7 9.3	July 10.5	+5 16 48.7
9.5	5 14 28.1	11.0	5 12 9.7
10.0	+5 17 38.3	11.5	+5 3 52.8

Ans. July 10^d 3^h $27^m.4$.

4. Given the equation

$$\sin(z-43°) = 0.92 \sin^4 z$$

to determine the root which falls in the second quadrant.

Ans. 101° 17′ 43″.

5. Given the following table of the longitude of *Mercury's* ascending node (θ):

Year	θ
	° ′ ″
1700	44 46 34.42
1800	45 57 39.28
1900	47 8 45.40
2000	48 19 52.78
2100	49 31 1.42

Express θ as a function of τ; where τ is the elapsed time from 1900, reckoned in terms of one century as the unit.

Ans. $\theta = 47° \; 8' \; 45''.40 + 4266''.75\tau + 0''.630\tau^2$.

APPENDIX.

ON THE SYMBOLIC METHOD OF DEVELOPMENT.

89. While many of the formulae and results in the foregoing text have been derived by somewhat indirect methods, yet the processes employed in every case have involved nothing but purely algebraic operations and principles.

For the benefit of such students as may be interested, we shall now devote a brief space to the more direct and potent form of development known as the *symbolic method*. In this our only purpose is to exhibit the simple manner in which the fundamental formulae of the text may be deduced; leaving the student to enter for himself upon the broader field thus opened by suggestion.

90. Let us define the *symbol of operation* Δ by the relation

$$\Delta F(T) = F(T+\omega) - F(T) \tag{368}$$

from which we formulate the following

Definition: *The operation of Δ upon any function of T produces the increment in the function which corresponds to the finite increment ω in the variable, T.*

The relation (368) may be more briefly expressed in the form

$$\Delta F_n = F_{n+1} - F_n = \Delta'_n \tag{369}$$

where n can have any value. Thus, taking $n = 0$, and referring to the schedule on page 15, we have

$$\Delta F_0 = F_1 - F_0 = \Delta'_0 \tag{370}$$

Similarly

$$\left.\begin{aligned} \Delta F_1 &= F_2 - F_1 = \Delta'_1 \\ \Delta F_2 &= F_3 - F_2 = \Delta'_2 \\ &\cdots\cdots\cdots\cdots \\ \Delta F_s &= F_{s+1} - F_s = \Delta'_s \end{aligned}\right\} \tag{371}$$

Thus it is evident that the effect of operating with Δ upon any *tabular* function is simply to form the *first difference* of that function and the succeeding tabular value. Whence it is evident that we have

$$\left.\begin{array}{l} \Delta\Delta F_0 = \Delta(\varDelta'_0) = \varDelta''_0 \\ \Delta\Delta F_1 = \Delta(\varDelta'_1) = \varDelta''_1 \\ \cdots\cdots\cdots\cdots \\ \Delta\Delta F_s = \Delta(\varDelta'_s) = \varDelta''_s \end{array}\right\} \quad (372)$$

It follows that the operation of $\Delta\Delta$ upon any tabular function produces the second difference bearing the same subscript. But this double operation of Δ may be conveniently characterized by Δ^2; hence we write

$$\Delta^2 F_0 = \varDelta''_0 \quad , \quad \Delta^2 F_1 = \varDelta''_1 \quad , \quad\quad , \quad \Delta^2 F_s = \varDelta''_s \quad (373)$$

In like manner, i denoting *any* integer, we have

$$\left.\begin{array}{l} \Delta^i F_0 = \Delta(\Delta^{i-1} F_0) = \Delta(\varDelta_0^{(i-1)}) = \varDelta_0^{(i)} \\ \Delta^i F_1 = \Delta(\Delta^{i-1} F_1) = \Delta(\varDelta_1^{(i-1)}) = \varDelta_1^{(i)} \\ \cdots\cdots\cdots\cdots\cdots\cdots \\ \Delta^i F_s = \Delta(\Delta^{i-1} F_s) = \Delta(\varDelta_s^{(i-1)}) = \varDelta_s^{(i)} \end{array}\right\} \quad (374)$$

and, more generally, n being a non-integer,

$$\Delta^i F_n = (\Delta\Delta\Delta \ldots . \; i \text{ times})\, F_n = \varDelta_n^{(i)} \quad (375)$$

91. Let us now consider the operation of differentiating $F(T)$ with respect to T and multiplying the derivative by ω. Denoting the operator in this process by D, we then have

$$\mathsf{D}F_n = \omega \frac{dF_n}{dT} = \omega F'_n \quad (376)$$

also

$$\mathsf{D}^2 F_n = \mathsf{DD}F_n = \omega \frac{d}{dT}(\omega F'_n) = \omega^2 F''_n \quad (377)$$

$$\mathsf{D}^i F_n = (\mathsf{DDD} \ldots . \; i \text{ times})\, F_n = \left(\omega \frac{d}{dT}\right)^i F_n = \omega^i F_n^{(i)} \quad (378)$$

92. The fundamental laws or principles governing the combination of symbols of *quantity* in algebraic operations are the following:

I. The *Distributive Law*, by virtue of which

$$a(p+q+r) = ap + aq + ar$$

II. The *Commutative Law*, expressed by the equation

$$ab = ba$$

III. The *Index Law*, which asserts the relation

$$a^r \times a^s = a^{r+s}$$

We proceed to show that the symbols of *operation*, Δ and D, when combined each with itself or with symbols of *quantity* in the manner indicated below, also obey these fundamental laws; and hence that, wherever found in similar combinations, Δ *and* D *may be treated algebraically precisely as if they were themselves mere symbols of quantity.* We shall first consider the symbol Δ.

(1). By definition, we have

$$\begin{aligned}\Delta(F_n+f_n+\ldots\ldots) &= (F_{n+1}+f_{n+1}+\ldots\ldots)-(F_n+f_n+\ldots\ldots)\\ &= (F_{n+1}-F_n)+(f_{n+1}-f_n)+\ldots\ldots\\ &= \Delta F_n+\Delta f_n+\ldots\ldots\end{aligned}$$

which proves the *Distributive Law* for the symbol Δ.

(2) The factor a being a constant, we have

$$\Delta aF_n = aF_{n+1} - aF_n = a(F_{n+1}-F_n) = a\Delta F_n$$

thus showing that Δ combines with *constant quantities* in accordance with the *Commutative Law*.

(3) r and s denoting positive integers, the relation (375) gives

$$\Delta^r\Delta^s F_n = \Delta^r(\Delta^s F_n) = \Delta^r \varDelta_n^{(s)} = \varDelta_n^{(r+s)} = \Delta^{r+s}F_n$$

or

$$\Delta^r\Delta^s = \Delta^{r+s}$$

Therefore, so far as *positive integral indices* are concerned, the symbol Δ obeys the *Index Law*.

93. Retaining the limitations and the notation used above, similar results are easily obtained for the operator D, as follows:

$$(1) \qquad \mathsf{D}(F_n+f_n+\ldots.) = \omega\frac{d}{dT}(F_n+f_n+\ldots.)$$
$$= \omega\frac{dF_n}{dT}+\omega\frac{df_n}{dT}+\ldots.$$
$$= \mathsf{D}F_n+\mathsf{D}f_n+\ldots.$$

$$(2) \qquad \mathsf{D}aF_n = \left(\omega\frac{d}{dT}\right)aF_n = a\omega\frac{dF_n}{dT} = a\mathsf{D}F_n$$

$$(3) \qquad \mathsf{D}^r\mathsf{D}^sF_n = \left(\omega^r\frac{d^r}{dT^r}\right)\left(\omega^s\frac{d^s}{dT^s}\right)F_n = \omega^r\omega^s\left(\frac{d^r}{dT^r}\right)\left(\frac{d^s}{dT^s}\right)F_n$$
$$= \omega^{r+s}\left(\frac{d^{r+s}}{dT^{r+s}}\right)F_n = \mathsf{D}^{r+s}F_n$$

These relations prove that — within the limitations imposed — the symbol D obeys the fundamental laws of algebraic combination.

94. To a limited extent it is necessary to consider negative powers of Δ and D. Now the meaning and use of $\Delta^{-1}, \Delta^{-2}, \ldots.$, and of $\mathsf{D}^{-1}, \mathsf{D}^{-2}, \ldots.$ are easily understood: thus, from the foregoing definitions, we have

$$\Delta('F_n) = F_n$$

where $'F_n$ is defined as in the schedule on page 134. Then, in analogy with the usual mode of expressing inverse functions, we may write

$$'F_n = \Delta^{-1}F_n$$

Whence we have

$$\Delta\Delta^{-1}F_n = \Delta('F_n) = F_n \qquad (379)$$

which shows (1) that the operation of $\Delta\Delta^{-1}(=\Delta^0)$ leaves the subject function unaltered, and (2) that *negative powers of Δ also obey the Index Law.*

The relation

$$\Delta^{-1}F_n = {}'F_n \qquad (380)$$

may be taken as the definition of the operator Δ^{-1}. Similarly, we have

$$\Delta^{-2}F_n = {}''F_n \quad , \quad \Delta^{-3}F_n = {}'''F_n \quad , \qquad (381)$$

Again, consider the relation

$$\mathsf{D}F_n = \omega\frac{dF_n}{dT} = v \qquad (382)$$

which, from the point of view above taken, may be written

$$F_n = \mathsf{D}^{-1}v \qquad (383)$$

Then we have

$$\mathsf{D}\mathsf{D}^{-1}v = \mathsf{D}F_n = v \tag{384}$$

whence we see that *negative* powers of D likewise follow the *Index Law*.

Moreover, from equation (382), we obtain

$$dF_n = \omega^{-1}v\,dT$$

and therefore

$$F_n = \omega^{-1}\int v\,dT$$

which, with (383), gives

$$\mathsf{D}^{-1}v = \omega^{-1}\int v\,dT \tag{385}$$

It follows that the operation of D^{-1} is equivalent to an integration. More specifically: *Operating upon any function with* D^{-1} *integrates that function with respect to* T *and divides the resulting integral by* ω.

In like manner we have

$$\mathsf{D}^{-2}F_n = \omega^{-2}\int\int F_n\,dT^2 \tag{386}$$

and so on.

95. Having thus defined and explained the use of the symbols of operation, $\Delta^{-2}, \Delta^{-1}, \Delta^0, \Delta, \Delta^2, \ldots\ldots$, and $\mathsf{D}^{-2}, \mathsf{D}^{-1}, \mathsf{D}^0, \mathsf{D}, \mathsf{D}^2, \ldots\ldots$; and having shown that these symbols may in general be combined algebraically as if they were merely symbols of quantity, we now proceed to derive the fundamental relations of the text, as originally proposed.

96. The theorem of the change in sign of the odd orders of differences caused by inverting a given series of functions is easily proved. To this end, let us suppose that $\varDelta_i^{(r)}$, of the direct or given series, becomes $[\varDelta_i^{(r)}]$ when that series has been inverted. Then, since

$$\Delta F_i = F_{i+1} - F_i = \varDelta_i'$$

we have

$$-\Delta F_i = F_i - F_{i+1} = [\varDelta_i']$$

Whence, regarding $-\Delta$ as operator, it follows that

$$(-\Delta)^2F_i = [\varDelta_i''], \qquad (-\Delta)^3F_i = [\varDelta_i'''], \qquad \ldots\ldots, \qquad (-\Delta)^rF_i = [\varDelta_i^{(r)}]$$

and therefore

$$[\varDelta_i^{(r)}] = (-\Delta)^rF_i = (-1)^r\Delta^rF_i = (-1)^r\varDelta_i^{(r)} \tag{387}$$

which establishes Theorem III.

97. By definition, we have

$$\Delta F_n = F_{n+1} - F_n$$

hence

$$(1+\Delta)F_n = F_n + \Delta F_n = F_{n+1} = F(\overline{t+n\omega}+\omega)$$

$$= F_n + \omega F'_n + \frac{\omega^2}{2!} F''_n + \frac{\omega^3}{3!} F'''_n + \ldots.$$

$$= F_n + \mathrm{D}F_n + \frac{\mathrm{D}^2}{2!} F_n + \frac{\mathrm{D}^3}{3!} F_n + \ldots.$$

$$= \left(1 + \mathrm{D} + \frac{\mathrm{D}^2}{2!} + \frac{\mathrm{D}^3}{3!} + \ldots.\right) F_n = e^{\mathrm{D}} F_n$$

where e is the base of the natural system of logarithms. We have, therefore,

$$1 + \Delta = e^{\mathrm{D}} \tag{388}$$

which is the fundamental relation between Δ and D.

98. From (388), we get

$$\Delta = e^{\mathrm{D}} - 1 = \mathrm{D} + \frac{\mathrm{D}^2}{2!} + \frac{\mathrm{D}^3}{3!} + \frac{\mathrm{D}^4}{4!} + \ldots. \tag{389}$$

and hence, by involution,

$$\left.\begin{aligned} \Delta^2 &= \mathrm{D}^2 + \mathrm{D}^3 + \tfrac{7}{12}\mathrm{D}^4 + \tfrac{1}{4}\mathrm{D}^5 + \ldots. \\ \Delta^3 &= \mathrm{D}^3 + \tfrac{3}{2}\mathrm{D}^4 + \tfrac{5}{4}\mathrm{D}^5 + \tfrac{3}{4}\mathrm{D}^6 + \ldots. \\ &\ldots\ldots\ldots\ldots\ldots\ldots \\ \Delta^i &= \mathrm{D}^i + \tfrac{i}{2}\mathrm{D}^{i+1} + \tfrac{i}{24}(3i+1)\mathrm{D}^{i+2} + \ldots. \end{aligned}\right\} \tag{390}$$

These expressions are equivalent to the formulae (21).

Again, from the last of (390), we derive

$$\Delta^i F_t = (\mathrm{D}^i + a_1\mathrm{D}^{i+1} + a_2\mathrm{D}^{i+2} + \ldots.)F_t$$

that is

$$\Delta_t^{(i)} = \omega^i F_t^{(i)} + a_1\omega^{i+1}F_t^{(i+1)} + a_2\omega^{i+2}F_t^{(i+2)} + \ldots. \tag{391}$$

where for brevity we have written $a_1, a_2, \ldots.$ to denote the coefficients of $\mathrm{D}^{i+1}, \mathrm{D}^{i+2}, \ldots.$ in (390). Whence, if $F(T) \equiv aT^i + \beta T^{i-1} + \gamma T^{i-2} + \ldots.$, we have

$$\Delta_t^{(i)} = \omega^i F_t^{(i)} = a\omega^i \frac{d^i}{dT^i}(T^i) = a\omega^i\, i!$$

which is the algebraic statement of Theorem V.

99. Expressing the relation (388) in logarithmic form, we get

$$\mathsf{D} = \log_e(1+\Delta) = \Delta - \frac{\Delta^2}{2} + \frac{\Delta^3}{3} - \frac{\Delta^4}{4} + \dots \tag{392}$$

whence

$$\left.\begin{aligned} \mathsf{D}^2 &= \Delta^2 - \Delta^3 + \tfrac{11}{12}\Delta^4 - \tfrac{5}{6}\Delta^5 + \dots \\ \mathsf{D}^3 &= \Delta^3 - \tfrac{3}{2}\Delta^4 + \tfrac{7}{4}\Delta^5 - \dots \\ &\dots\dots\dots\dots \end{aligned}\right\} \tag{393}$$

From these relations the formulae (45) — or the equivalent group (165) — immediately follow.

100. We next consider the question of reducing the tabular interval from ω to $m\omega$, as discussed in § 19. Since in the preceding definitions of Δ and D the magnitude of the interval is arbitrary, we have here only to denote by ∂ and d the corresponding symbols in the reduced series; evidently the same relations will then exist between ∂ and d as were found above for Δ and D. Thus we obtain

$$\mathsf{d} = m\omega\frac{d}{dT} = m\left(\omega\frac{d}{dT}\right) = m\mathsf{D} \tag{394}$$

and since, by (388), we have

$$1 + \Delta = e^{\mathsf{D}}$$

we must have also

$$1 + \partial = e^{\mathsf{d}} = e^{m\mathsf{D}} \tag{395}$$

Whence we find

$$1 + \partial = (1+\Delta)^m = 1 + m\Delta + \frac{m(m-1)}{|\underline{2}}\Delta^2 + \frac{m(m-1)(m-2)}{|\underline{3}}\Delta^3 + \dots$$

and therefore

$$\left.\begin{aligned} \partial &= m\Delta + \frac{m(m-1)}{|\underline{2}}\Delta^2 + \frac{m(m-1)(m-2)}{|\underline{3}}\Delta^3 + \dots \\ \partial^2 &= m^2\Delta^2 + m^2(m-1)\Delta^3 + \dots\dots \\ \partial^3 &= m^3\Delta^3 + \dots\dots \\ &\dots\dots\dots\dots \end{aligned}\right\} \tag{396}$$

which are equivalent to the relations expressed in (64).

101. The equation

$$\Delta F_0 = F_1 - F_0$$

may be written in the form

$$(1+\Delta)F_0 = F_1 \tag{397}$$

Hence the binomial $1+\Delta$ may be defined as an operator whose effect is to raise by unity the subscript of the subject function. Whence we have

$$\left.\begin{aligned}(1+\Delta)^2 F_0 &= (1+\Delta)\,F_1 = F_2\\(1+\Delta)^3 F_0 &= (1+\Delta)\,F_2 = F_3\end{aligned}\right\} \qquad (398)$$

and generally

$$(1+\Delta)^n F_0 = F_n \qquad (399)$$

We therefore obtain

$$F_n = (1+\Delta)^n F_0 = \left(1+n\Delta+\frac{n\,(n-1)}{\underline{|2}}\,\Delta^2+\frac{n\,(n-1)(n-2)}{\underline{|3}}\,\Delta^3+\ldots\ldots\right)F_0$$

or

$$F_n = F_0 + n\Delta'_0+\frac{n\,(n-1)}{\underline{|2}}\,\Delta''_0+\frac{n\,(n-1)(n-2)}{\underline{|3}}\,\Delta'''_0+\ldots\ldots \qquad (400)$$

which is the fundamental formula of interpolation due to NEWTON.

102. We now find it convenient to introduce a new symbol of operation, which, from its similarity and relation to Δ, we shall designate ∇: this operator is defined by the equation

$$\nabla F_i = F_i - F_{i-1} = \Delta'_{i-1} \qquad (401)$$

From this relation we at once derive

$$\left.\begin{aligned}\nabla^2 F_i &= \nabla\Delta'_{i-1} = \Delta''_{i-2}\\\nabla^3 F_i &= \nabla\Delta''_{i-2} = \Delta'''_{i-3}\\\nabla^4 F_i &= \nabla\Delta'''_{i-3} = \Delta^{\text{iv}}_{i-4}\\&\ldots\ldots\ldots\ldots\end{aligned}\right\} \qquad (402)$$

whence it appears that the operation of ∇^r upon any tabular function produces the difference of order r which falls upon the *upward inclined diagonal* through that function; whereas the successive operations of Δ produce, as already shown, those differences falling upon the *downward* diagonal line. Moreover, from the complete similarity of character of these two operators, it is obvious that ∇ likewise follows the fundamental laws of algebraic combination.

The relation between ∇ and Δ is easily found: thus, from (401), we obtain

$$(1-\nabla)\,F_i = F_{i-1} \qquad (403)$$

also, from (397), we have

$$(1+\Delta)\,F_{i-1} = F_i \qquad (404)$$

Whence we find

$$(1+\Delta)(1-\nabla)F_i = (1+\Delta)F_{i-1} = F_i$$

and therefore

$$1-\nabla = (1+\Delta)^{-1} \tag{405}$$

which gives

$$\log(1-\nabla) = -\log(1+\Delta) \tag{406}$$

Again, combining (388) and (405), we obtain

$$1-\nabla = e^{-D} \tag{407}$$

103. As an immediate application of the preceding relations, let us derive the formula (75). By means of (388), equation (399) becomes

$$F_n = (1+\Delta)^n F_0 = e^{nD} F_0$$

whence, changing the sign of n, we find

$$\begin{aligned} F_{-n} &= e^{-nD}F_0 = (e^{-D})^n F_0 = (1-\nabla)^n F_0 \\ &= \left(1 - n\nabla + \frac{n(n-1)}{\underline{|2}}\nabla^2 - \frac{n(n-1)(n-2)}{\underline{|3}}\nabla^3 + \ldots\right) F_0 \end{aligned}$$

Therefore

$$F_{-n} = F_0 - nA'_{-1} + \frac{n(n-1)}{\underline{|2}}A''_{-2} - \frac{n(n-1)(n-2)}{\underline{|3}}A'''_{-3} + \ldots \tag{408}$$

which is Newton's Formula for *backward* interpolation, as given by (75).

104. Formula (66) of the text is easily deduced by means of the identity

$$\Delta = (1+\Delta) - 1$$

Thus we find

$$\begin{aligned} \Delta^i F_0 &= \{(1+\Delta)-1\}^i F_0 \\ &= \left\{(1+\Delta)^i - i(1+\Delta)^{i-1} + \frac{i(i-1)}{\underline{|2}}(1+\Delta)^{i-2} - \ldots \right\} F_0 \end{aligned}$$

whence, by (399), we obtain

$$A_0^{(i)} = F_i - iF_{i-1} + \frac{i(i-1)}{\underline{|2}}F_{i-2} - \frac{i(i-1)(i-2)}{\underline{|3}}F_{i-3} + \ldots \tag{409}$$

which is the same as equation (66).

105. We now pass to the derivation of the fundamental formulae of mechanical quadrature. Since $\mathsf{D} = \log(1+\triangle)$, we have

$$\mathsf{D}^{-1}F_n = \{\log(1+\triangle)\}^{-1}F_n = \left(\triangle - \frac{\triangle^2}{2} + \frac{\triangle^3}{3} - \frac{\triangle^4}{4} + \ldots\ldots\right)^{-1}F_n$$
$$= (\triangle^{-1} + \tfrac{1}{2} - \tfrac{1}{12}\triangle + \tfrac{1}{24}\triangle^2 - \tfrac{19}{720}\triangle^3 + \tfrac{3}{160}\triangle^4 - \tfrac{863}{60480}\triangle^5 + \ldots\ldots)F_n$$

Whence, interpreting the first member according to (385), and the term $\triangle^{-1}F_n$ as in (380), we find

$$\omega^{-1}\int F_n dT = {}'F_n + \tfrac{1}{2}F_n - \tfrac{1}{12}\Delta'_n + \tfrac{1}{24}\Delta''_n - \tfrac{19}{720}\Delta'''_n + \tfrac{3}{160}\Delta^{\text{iv}}_n - \tfrac{863}{60480}\Delta^{\text{v}}_n + \ldots \quad (410)$$

This is the fundamental relation of quadrature, from which the formula (a) of (250) is at once derived. To obtain (b) of (250) involving the differences $\Delta'_{n-1}, \Delta''_{n-2}, \Delta'''_{n-3}, \ldots\ldots,$ we have only to employ the relation (406), and the above development becomes

$$\mathsf{D}^{-1}F_n = \{\log(1+\triangle)\}^{-1}F_n = \{-\log(1-\nabla)\}^{-1}F_n$$
$$= \left(\nabla + \frac{\nabla^2}{2} + \frac{\nabla^3}{3} + \frac{\nabla^4}{4} + \frac{\nabla^5}{5} + \ldots\ldots\right)^{-1}F_n$$
$$= (\nabla^{-1} - \tfrac{1}{2} - \tfrac{1}{12}\nabla - \tfrac{1}{24}\nabla^2 - \tfrac{19}{720}\nabla^3 - \tfrac{3}{160}\nabla^4 - \tfrac{863}{60480}\nabla^5 - \ldots\ldots)F_n$$

the interpretation of which gives

$$\omega^{-1}\int F_n dT = {}'F_{n+1} - \tfrac{1}{2}F_n - \tfrac{1}{12}\Delta'_{n-1} - \tfrac{1}{24}\Delta''_{n-2} - \tfrac{19}{720}\Delta'''_{n-3} - \tfrac{3}{160}\Delta^{\text{iv}}_{n-4} - \tfrac{863}{60480}\Delta^{\text{v}}_{n-5} - \ldots \quad (411)$$

agreeing with formula (b) of (250).

106. Similarly, we obtain for the second integration

$$\mathsf{D}^{-2}F_n = \{\log(1+\triangle)\}^{-2}F_n = \left(\triangle - \frac{\triangle^2}{2} + \frac{\triangle^3}{3} - \frac{\triangle^4}{4} + \ldots\ldots\right)^{-2}F_n$$
$$= (\triangle^{-2} + \triangle^{-1} + \tfrac{1}{12} - \tfrac{1}{240}\triangle^2 + \tfrac{1}{240}\triangle^3 - \tfrac{221}{60480}\triangle^4 + \tfrac{19}{6048}\triangle^5 - \ldots)F_n$$

Now the first pair of terms in the right-hand member may be written

$$(\triangle^{-2} + \triangle^{-1})F_n = \triangle^{-2}(1+\triangle)F_n = \triangle^{-2}F_{n+1} = {}''F_{n+1}$$

and therefore the preceding expression becomes

$$\omega^{-2}\int\int F_n dT^2 = {}''F_{n+1} + \tfrac{1}{12}F_n - \tfrac{1}{240}\Delta''_n + \tfrac{1}{240}\Delta'''_n - \tfrac{221}{60480}\Delta^{\text{iv}}_n + \tfrac{19}{6048}\Delta^{\text{v}}_n - \ldots \quad (412)$$

from which (324) immediately follows.

Again, we find

$$\begin{aligned} D^{-2}F_n &= \{\log(1+\Delta)\}^{-2}F_n = \{-\log(1-\nabla)\}^{-2}F_n \\ &= \left(\nabla+\frac{\nabla^2}{2}+\frac{\nabla^3}{3}+\frac{\nabla^4}{4}+\frac{\nabla^5}{5}+\ \ldots\ldots\right)^{-2}F_n \\ &= (\nabla^{-2}-\nabla^{-1}+\tfrac{1}{12}-\tfrac{1}{240}\nabla^2-\tfrac{1}{240}\nabla^3-\tfrac{221}{60480}\nabla^4-\tfrac{19}{6048}\nabla^6-\ .\ .\ )F_n \end{aligned} \quad (413)$$

Transforming the first two terms of the last expression, we obtain

$$(\nabla^{-2}-\nabla^{-1})F_n = \nabla^{-2}(1-\nabla)F_n = \nabla^{-2}(1+\Delta)^{-1}F_n$$

Now, because the operation of $1+\Delta$ raises by unity the subscript of the subject function (§ 101), it follows that the operation of $(1+\Delta)^{-1}$ *diminishes* that subscript by one unit. Accordingly, we have

$$(\nabla^{-2}-\nabla^{-1})F_n = \nabla^{-2}(1+\Delta)^{-1}F_n = \nabla^{-2}F_{n-1} = {}^{\prime\prime}F_{n+1}$$

and hence the relation (413) gives

$$\omega^{-2}\int\!\!\int F_n dT^2 = {}^{\prime\prime}F_{n+1}+\tfrac{1}{12}F_n-\tfrac{1}{240}\Delta''_{n-2}-\tfrac{1}{240}\Delta'''_{n-3}-\tfrac{221}{60480}\Delta^{\text{iv}}_{n-4}-\tfrac{19}{6048}\Delta^{\text{v}}_{n-5}-\ .\ .\ . \quad (414)$$

which is equivalent to the formula (326). These expressions complete the fundamental relations of mechanical quadrature.

TABLES.

Interval	Binomial Coefficients for							
	Δ''		Δ'''		Δiv		Δv	
n	$\frac{n(n-1)}{2}$	Diff.	$\frac{n(n-1)(n-2)}{6}$	Diff.	$\frac{n\dots(n-3)}{24}$	Diff.	$\frac{n\dots(n-4)}{120}$	Diff.
0.00	.00000	—495	.00000	+328	.00000	—245	.00000	+196
.01	—.00495	485	+.00328	319	—.00245	237	+.00196	188
.02	.00980	475	.00647	308	.00482	227	.00384	179
.03	.01455	465	.00955	299	.00709	219	.00563	172
.04	.01920	455	.01254	290	.00928	211	.00735	164
.05	.02375	445	.01544	280	.01139	201	.00899	157
.06	—.02820	435	+.01824	270	—.01340	194	+.01056	150
.07	.03255	425	.02094	261	.01534	185	.01206	142
.08	.03680	415	.02355	252	.01719	178	.01348	135
.09	.04095	405	.02607	243	.01897	169	.01483	129
.10	.04500	395	.02850	234	.02066	162	.01612	121
.11	—.04895	385	+.03084	225	—.02228	154	+.01733	116
.12	.05280	375	.03309	216	.02382	147	.01849	109
.13	.05655	365	.03525	207	.02529	140	.01958	102
.14	.06020	355	.03732	199	.02669	132	.02060	97
.15	.06375	345	.03931	191	.02801	125	.02157	90
.16	—.06720	335	+.04122	182	—.02926	119	+.02247	85
.17	.07055	325	.04304	173	.03045	111	.02332	80
.18	.07380	315	.04477	166	.03156	105	.02412	73
.19	.07695	305	.04643	157	.03261	99	.02485	69
.20	.08000	295	.04800	149	.03360	92	.02554	63
.21	—.08295	285	+.04949	142	—.03452	86	+.02617	58
.22	.08580	275	.05091	133	.03538	80	.02675	53
.23	.08855	265	.05224	126	.03618	74	.02728	48
.24	.09120	—255	.05350	+119	.03692	— 68	.02776	+ 44
0.25	—.09375		+.05469		—.03760		+.02820	

Interval	Binomial Coefficients for							
	Δ''		Δ'''		Δiv		Δv	
n	$\frac{n(n-1)}{2}$	Diff.	$\frac{n(n-1)(n-2)}{6}$	Diff.	$\frac{n\dots(n-3)}{24}$	Diff.	$\frac{n\dots(n-4)}{120}$	Diff.
0.25	—.09375	—245	+.05469	+111	—.03760	—62	+.02820	+39
.26	.09620	235	.05580	103	.03822	57	.02859	34
.27	.09855	225	.05683	96	.03879	51	.02893	31
.28	.10080	215	.05779	89	.03930	46	.02924	26
.29	.10295	205	.05868	82	.03976	40	.02950	22
.30	.10500	195	.05950	75	.04016	36	.02972	18
.31	—.10695	185	+.06025	68	—.04052	30	+.02990	14
.32	.10880	175	.06093	61	.04082	26	.03004	11
.33	.11055	165	.06154	54	.04108	21	.03015	7
.34	.11220	155	.06208	48	.04129	16	.03022	+ 4
.35	.11375	145	.06256	42	.04145	11	.03026	0
.36	—.11520	135	+.06298	35	—.04156	8	+.03026	— 3
.37	.11655	125	.06333	28	.04164	— 3	.03023	6
.38	.11780	115	.06361	23	.04167	+ 2	.03017	10
.39	.11895	105	.06384	16	.04165	5	.03007	12
.40	.12000	95	.06400	10	.04160	9	.02995	15
.41	—.12095	85	+.06410	+ 5	—.04151	13	+.02980	18
.42	.12180	75	.06415	— 2	.04138	17	.02962	20
.43	.12255	65	.06413	7	.04121	21	.02942	23
.44	.12320	55	.06406	12	.04100	24	.02919	25
.45	.12375	45	.06394	18	.04076	27	.02894	28
.46	—.12420	35	+.06376	24	—.04049	31	+.02866	30
.47	.12455	25	.06352	29	.04018	34	.02836	32
.48	.12480	15	.06323	34	.03984	38	.02804	34
.49	.12495	— 5	.06289	— 39	.03946	+40	.02770	—36
0.50	—.12500		+.06250		—.03906		+.02734	

Interval	Binomial Coefficients for							
	Δ''		Δ'''		Δ^{iv}		Δ^{v}	
n	$\frac{n(n-1)}{2}$	Diff.	$\frac{n(n-1)(n-2)}{6}$	Diff.	$\frac{n\ldots(n-3)}{24}$	Diff.	$\frac{n\ldots(n-4)}{120}$	Diff.
0.50	−.12500	+ 5	+.06250	− 44	−.03906	+43	+.02734	−38
.51	.12495	15	.06206	49	.03863	46	.02696	39
.52	.12480	25	.06157	54	.03817	48	.02657	42
.53	.12455	35	.06103	59	.03769	52	.02615	43
.54	.12420	45	.06044	63	.03717	53	.02572	44
.55	.12375	55	.05981	67	.03664	57	.02528	46
.56	−.12320	65	+.05914	72	−.03607	58	+.02482	48
.57	.12255	75	.05842	77	.03549	61	.02434	48
.58	.12180	85	.05765	80	.03488	63	.02386	50
.59	.12095	95	.05685	85	.03425	65	.02336	51
.60	.12000	105	.05600	89	.03360	67	.02285	52
.61	−.11895	115	+.05511	92	−.03293	69	+.02233	53
.62	.11780	125	.05419	97	.03224	70	.02180	55
.63	.11655	135	.05322	100	.03154	73	.02125	54
.64	.11520	145	.05222	103	.03081	74	.02071	56
.65	.11375	155	.05119	107	.03007	75	.02015	57
.66	−.11220	165	+.05012	111	−.02932	77	+.01958	57
.67	.11055	175	.04901	114	.02855	78	.01901	57
.68	.10880	185	.04787	117	.02777	80	.01844	59
.69	.10695	195	.04670	120	.02697	81	.01785	58
.70	.10500	205	.04550	123	.02616	82	.01727	59
.71	−.10295	215	+.04427	126	−.02534	83	+.01668	60
.72	.10080	225	.04301	129	.02451	83	.01608	60
.73	.09855	235	.04172	132	.02368	85	.01548	60
.74	.09620	+245	.04040	−134	.02283	+86	.01488	−60
0.75	−.09375		+.03906		−.02197		+.01428	

Interval	Binomial Coefficients for							
	Δ''		Δ'''		Δ^{iv}		Δ^{v}	
n	$\frac{n(n-1)}{2}$	Diff.	$\frac{n(n-1)(n-2)}{6}$	Diff.	$\frac{n\ldots(n-3)}{24}$	Diff.	$\frac{n\ldots(n-4)}{120}$	Diff.
0.75	−.09375	+255	+.03906	−136	−.02197	+86	+.01428	−60
.76	.09120	265	.03770	139	.02111	87	.01368	60
.77	.08855	275	.03631	142	.02024	87	.01308	61
.78	.08580	285	.03489	143	.01937	89	.01247	60
.79	.08295	295	.03346	146	.01848	88	.01187	61
.80	.08000	305	.03200	148	.01760	89	.01126	60
.81	−.07695	315	+.03052	149	−.01671	89	+.01066	60
.82	.07380	325	.02903	152	.01582	89	.01006	60
.83	.07055	335	.02751	153	.01493	90	.00946	59
.84	.06720	345	.02598	154	.01403	89	.00887	59
.85	.06375	355	.02444	156	.01314	90	.00828	59
.86	−.06020	365	+.02288	158	−.01224	90	+.00769	59
.87	.05655	375	.02130	159	.01134	89	.00710	58
.88	.05280	385	.01971	160	.01045	90	.00652	58
.89	.04895	395	.01811	161	.00955	89	.00594	57
.90	.04500	405	.01650	162	.00866	89	.00537	57
.91	−.04095	415	+.01488	163	−.00777	88	+.00480	56
.92	.03680	425	.01325	164	.00689	88	.00424	55
.93	.03255	435	.01161	165	.00601	88	.00369	55
.94	.02820	445	.00996	165	.00513	87	.00314	54
.95	.02375	455	.00831	165	.00426	87	.00260	54
.96	−.01920	465	+.00666	166	−.00339	85	+.00206	52
.97	.01455	475	.00500	167	.00254	86	.00154	52
.98	.00980	485	.00333	166	.00168	84	.00102	52
.99	−.00495	+495	+.00167	−167	−.00084	+84	+.00050	−50
1.00	.00000		.00000		.00000		.00000	

Interval	Stirling's Coefficients for							
	Δ''		Δ'''		Δ^{iv}		Δ^{v}	
n	$\frac{n^2}{2}$	Diff.	$\frac{n(n^2-1)}{6}$	Diff.	$\frac{n^2(n^2-1)}{24}$	Diff.	$\frac{n(n^2-1)(n^2-4)}{120}$	Diff.
0.00	.00000	+ 5	.00000	−167	.00000	0	.00000	+ 33
.01	+.00005	15	−.00167	166	.00000	− 2	+.00033	34
.02	.00020	25	.00333	167	−.00002	2	.00067	33
.03	.00045	35	.00500	166	.00004	3	.00100	33
.04	.00080	45	.00666	165	.00007	3	.00133	33
.05	.00125	55	.00831	165	.00010	5	.00166	33
.06	+.00180	65	−.00996	165	−.00015	5	+.00199	33
.07	.00245	75	.01161	164	.00020	6	.00232	33
.08	.00320	85	.01325	163	.00026	7	.00265	32
.09	.00405	95	.01488	162	.00033	8	.00297	32
.10	.00500	105	.01650	161	.00041	9	.00329	32
.11	+.00605	115	−.01811	160	−.00050	9	+.00361	32
.12	.00720	125	.01971	159	.00059	10	.00393	31
.13	.00845	135	.02130	158	.00069	11	.00424	31
.14	.00980	145	.02288	156	.00080	12	.00455	31
.15	.01125	155	.02444	154	.00092	12	.00486	30
.16	+.01280	165	−.02598	153	−.00104	13	+.00516	30
.17	.01445	175	.02751	152	.00117	14	.00546	30
.18	.01620	185	.02903	149	.00131	14	.00576	29
.19	.01805	195	.03052	148	.00145	15	.00605	29
.20	.02000	205	.03200	146	.00160	16	.00634	28
.21	+.02205	215	−.03346	143	−.00176	16	+.00662	27
.22	.02420	225	.03489	142	.00192	17	.00689	28
.23	.02645	235	.03631	139	.00209	17	.00717	26
.24	.02880	+245	.03770	−136	.00226	−18	.00743	+26
0.25	+.03125		−.03906		−.00244		+.00769	

Interval	Stirling's Coefficients for							
	Δ''		Δ'''		Δ^{iv}		Δ^{v}	
n	$\frac{n^2}{2}$	Diff.	$\frac{n(n^2-1)}{6}$	Diff.	$\frac{n^2(n^2-1)}{24}$	Diff.	$\frac{n(n^2-1)(n^2-4)}{120}$	Diff.
0.25	+.03125	+255	−.03906	−134	−.00244	−19	+.00769	+25
.26	.03380	265	.04040	132	.00263	19	.00794	25
.27	.03645	275	.04172	129	.00282	19	.00819	24
.28	.03920	285	.04301	126	.00301	20	.00843	24
.29	.04205	295	.04427	123	.00321	20	.00867	23
.30	.04500	305	.04550	120	.00341	21	.00890	22
.31	+.04805	315	−.04670	117	−.00362	21	+.00912	21
.32	.05120	325	.04787	114	.00383	21	.00933	21
.33	.05445	335	.04901	111	.00404	22	.00954	19
.34	.05780	345	.05012	107	.00426	22	.00973	19
.35	.06125	355	.05119	103	.00448	22	.00992	19
.36	+.06480	365	−.05222	100	−.00470	22	+.01011	17
.37	.06845	375	.05322	97	.00492	23	.01028	17
.38	.07220	385	.05419	92	.00515	22	.01045	15
.39	.07605	395	.05511	89	.00537	23	.01060	15
.40	.08000	405	.05600	85	.00560	23	.01075	14
.41	+.08405	415	−.05685	80	−.00583	22	+.01089	13
.42	.08820	425	.05765	77	.00605	23	.01102	12
.43	.09245	435	.05842	72	.00628	22	.01114	11
.44	.09680	445	.05914	67	.00650	23	.01125	11
.45	.10125	455	.05981	63	.00673	22	.01136	9
.46	+.10580	465	−.06044	59	−.00695	22	+.01145	8
.47	.11045	475	.06103	54	.00717	22	.01153	7
.48	.11520	485	.06157	49	.00739	21	.01160	7
.49	.12005	+495	.06206	− 44	.00760	−21	.01167	+ 5
0.50	+.12500		−.06250		−.00781		+.01172	

Interval	Stirling's Coefficients for							
	Δ''		Δ'''		Δ^{iv}		Δ^{v}	
n	$\frac{n^2}{2}$	Diff.	$\frac{n(n^2-1)}{6}$	Diff.	$\frac{n^2(n^2-1)}{24}$	Diff.	$\frac{n(n^2-1)(n^2-4)}{120}$	Diff.
0.50	+.12500		−.06250		−.00781		+.01172	
		+505		−39		−21		+4
.51	.13005		.06289		.00802		.01176	
		515		34		20		3
.52	.13520		.06323		.00822		.01179	
		525		29		20		2
.53	.14045		.06352		.00842		.01181	
		535		24		19		+1
.54	.14580		.06376		.00861		.01182	
		545		18		18		0
.55	.15125		.06394		.00879		.01182	
		555		12		18		−1
.56	+.15680		−.06406		−.00897		+.01181	
		565		7		17		2
.57	.16245		.06413		.00914		.01179	
		575		−2		16		4
.58	.16820		.06415		.00930		.01175	
		585		+5		16		5
.59	.17405		.06410		.00946		.01170	
		595		10		14		5
.60	.18000		.06400		.00960		.01165	
		605		16		14		7
.61	+.18605		−.06384		−.00974		+.01158	
		615		23		12		8
.62	.19220		.06361		.00986		.01150	
		625		28		11		9
.63	.19845		.06333		.00997		.01141	
		635		35		11		10
.64	.20480		.06298		.01008		.01131	
		645		42		9		12
.65	.21125		.06256		.01017		.01119	
		655		48		7		13
.66	+.21780		−.06208		−.01024		+.01106	
		665		54		7		13
.67	.22445		.06154		.01031		.01093	
		675		61		5		15
.68	.23120		.06093		.01036		.01078	
		685		68		3		16
.69	.23805		.06025		.01039		.01062	
		695		75		2		18
.70	.24500		.05950		.01041		.01044	
		705		82		−1		18
.71	+.25205		−.05868		−.01042		+.01026	
		715		89		+2		20
.72	.25920		.05779		.01040		.01006	
		725		96		3		21
.73	.26645		.05683		.01037		.00985	
		735		103		5		22
.74	.27380		.05580		.01032		.00963	
		+745		+111		+7		−23
0.75	+.28125		−.05469		−.01025		+.00940	

Interval	Stirling's Coefficients for							
	Δ''		Δ'''		Δ^{iv}		Δ^{v}	
n	$\frac{n^2}{2}$	Diff.	$\frac{n(n^2-1)}{6}$	Diff.	$\frac{n^2(n^2-1)}{24}$	Diff.	$\frac{n(n^2-1)(n^2-4)}{120}$	Diff.
0.75	+.28125		−.05469		−.01025		+.00940	
		+755		+119		+8		−24
.76	.28880		.05350		.01017		.00916	
		765		126		11		26
.77	.29645		.05224		.01006		.00890	
		775		133		13		27
.78	.30420		.05091		.00993		.00863	
		785		142		16		28
.79	.31205		.04949		.00977		.00835	
		795		149		17		29
.80	.32000		.04800		.00960		.00806	
		805		157		20		30
.81	+.32805		−.04643		−.00940		+.00776	
		815		166		22		31
.82	.33620		.04477		.00918		.00745	
		825		173		25		33
.83	.34445		.04304		.00893		.00712	
		835		182		27		33
.84	.35280		.04122		.00866		.00679	
		845		191		31		35
.85	.36125		.03931		.00835		.00644	
		855		199		33		36
.86	+.36980		−.03732		−.00802		+.00608	
		865		207		35		36
.87	.37845		.03525		.00767		.00572	
		875		216		39		38
.88	.38720		.03309		.00728		.00534	
		885		225		42		39
.89	.39605		.03084		.00686		.00495	
		895		234		45		40
.90	.40500		.02850		.00641		.00455	
		905		243		48		42
.91	+.41405		−.02607		−.00593		+.00413	
		915		252		51		42
.92	.42320		.02355		.00542		.00371	
		925		261		55		43
.93	.43245		.02094		.00487		.00328	
		935		270		58		44
.94	.44180		.01824		.00429		.00284	
		945		280		62		45
.95	.45125		.01544		.00367		.00239	
		955		290		66		46
.96	+.46080		−.01254		−.00301		+.00193	
		965		299		69		47
.97	.47045		.00955		.00232		.00146	
		975		308		74		48
.98	.48020		.00647		.00158		.00098	
		985		319		77		48
.99	.49005		−.00328		−.00081		+.00050	
		+995		+328		+81		−50
1.00	+.50000		.00000		.00000		.00000	

Interval	Bessel's Coefficients for							
	J″		J‴		J^iv		J^v	
n	$\frac{n(n-1)}{2}$	Diff.	$\frac{n(n-1)(n-\frac{1}{2})}{6}$	Diff.	$\frac{(n+1)\ldots(n-2)}{24}$	Diff.	$\frac{(n+1)\ldots(n-\frac{1}{2})}{120}$	Diff.
0.00	.00000		.00000		.00000		.00000	
		−495		+81		+83		−8
.01	−.00495		+.00081		+.00083		−.00008	
		485		76		82		8
.02	.00980		.00157		.00165		.00016	
		475		71		81		7
.03	.01455		.00228		.00246		.00023	
		465		66		80		7
.04	.01920		.00294		.00326		.00030	
		455		62		79		6
.05	.02375		.00356		.00405		.00036	
		445		58		78		7
.06	−.02820		+.00414		+.00483		−.00043	
		435		53		77		5
.07	.03255		.00467		.00560		.00048	
		425		48		76		5
.08	.03680		.00515		.00636		.00053	
		415		45		74		5
.09	.04095		.00560		.00710		.00058	
		405		40		74		5
.10	.04500		.00600		.00784		.00063	
		395		36		72		4
.11	−.04895		+.00636		+.00856		−.00067	
		385		33		70		3
.12	.05280		.00669		.00926		.00070	
		375		28		70		4
.13	.05655		.00697		.00996		.00074	
		365		25		68		3
.14	.06020		.00722		.01064		.00077	
		355		22		66		2
.15	.06375		.00744		.01130		.00079	
		345		18		65		2
.16	−.06720		+.00762		+.01195		−.00081	
		335		14		64		2
.17	.07055		.00776		.01259		.00083	
		325		11		62		2
.18	.07380		.00787		.01321		.00085	
		315		8		60		−1
.19	.07695		.00795		.01381		.00086	
		305		5		59		0
.20	.08000		.00800		.01440		.00086	
		295		+ 2		57		−1
.21	−.08295		+.00802		+.01497		−.00087	
		285		− 1		56		0
.22	.08580		.00801		.01553		.00087	
		275		4		54		0
.23	.08855		.00797		.01607		.00087	
		265		7		52		+1
.24	.09120		.00790		.01659		.00086	
		−255		− 9		+50		+1
0.25	−.09375		+.00781		+.01709		−.00085	

Interval	Bessel's Coefficients for							
	J″		J‴		J^iv		J^v	
n	$\frac{n(n-1)}{2}$	Diff.	$\frac{n(n-1)(n-\frac{1}{2})}{6}$	Diff.	$\frac{(n+1)\ldots(n-2)}{24}$	Diff.	$\frac{(n+1)\ldots(n-\frac{1}{2})}{120}$	Diff.
0.25	−.09375		+.00781		+.01709		−.00085	
		−245		−11		+49		+1
.26	.09620		.00770		.01758		.00084	
		235		14		46		1
.27	.09855		.00756		.01804		.00083	
		225		17		45		2
.28	.10080		.00739		.01849		.00081	
		215		18		43		2
.29	.10295		.00721		.01892		.00079	
		205		21		42		2
.30	.10500		.00700		.01934		.00077	
		195		23		39		2
.31	−.10695		+.00677		+.01973		−.00075	
		185		24		38		3
.32	.10880		.00653		.02011		.00072	
		175		27		35		2
.33	.11055		.00626		.02046		.00070	
		165		28		34		3
.34	.11220		.00598		.02080		.00067	
		155		29		31		4
.35	.11375		.00569		.02111		.00063	
		145		31		30		3
.36	−.11520		+.00538		+.02141		−.00060	
		135		33		28		4
.37	.11655		.00505		.02169		.00056	
		125		34		26		3
.38	.11780		.00471		.02195		.00053	
		115		35		23		4
.39	.11895		.00436		.02218		.00049	
		105		36		22		4
.40	.12000		.00400		.02240		.00045	
		95		37		20		4
.41	−.12095		+.00363		+.02260		−.00041	
		85		38		17		5
.42	.12180		.00325		.02277		.00036	
		75		39		16		4
.43	.12255		.00286		.02293		.00032	
		65		40		13		4
.44	.12320		.00246		.02306		.00028	
		55		40		12		5
.45	.12375		.00206		.02318		.00023	
		45		40		9		4
.46	−.12420		+.00166		+.02327		−.00019	
		35		41		7		5
.47	.12455		.00125		.02334		.00014	
		25		42		6		5
.48	.12480		.00083		.02340		.00009	
		15		41		3		4
.49	.12495		+.00042		.02343		−.00005	
		− 5		−42		+ 1		+5
0.50	−.12500		.00000		+.02344		.00000	

Interval	Bessel's Coefficients for							
	J''		J'''		J^{iv}		J^{v}	
n	$\frac{n(n-1)}{2}$	Diff.	$\frac{n(n-1)(n-\frac{1}{2})}{6}$	Diff.	$\frac{(n+1)\ldots(n-2)}{24}$	Diff.	$\frac{(n+1)\ldots(n-\frac{1}{2})}{120}$	Diff.
0.50	−.12500		.00000		+.02344		.00000	
		+ 5		—42		— 1		+5
.51	.12495		−.00042		.02343		+.00005	
		15		41		3		4
.52	.12480		.00083		.02340		.00009	
		25		42		6		5
.53	.12455		.00125		.02334		.00014	
		35		41		7		5
.54	.12420		.00166		.02327		.00019	
		45		40		9		4
.55	.12375		.00206		.02318		.00023	
		55		40		12		5
.56	−.12320		−.00246		+.02306		+.00028	
		65		40		13		4
.57	.12255		.00286		.02293		.00032	
		75		39		16		4
.58	.12180		.00325		.02277		.00036	
		85		38		17		5
.59	.12095		.00363		.02260		.00041	
		95		37		20		4
.60	.12000		.00400		.02240		.00045	
		105		36		22		4
.61	−.11895		−.00436		+.02218		+.00049	
		115		35		23		4
.62	.11780		.00471		.02195		.00053	
		125		34		26		3
.63	.11655		.00505		.02169		.00056	
		135		33		28		4
.64	.11520		.00538		.02141		.00060	
		145		31		30		3
.65	.11375		.00569		.02111		.00063	
		155		29		31		4
.66	−.11220		−.00598		+.02080		+.00067	
		165		28		34		3
.67	.11055		.00626		.02046		.00070	
		175		27		35		2
.68	.10880		.00653		.02011		.00072	
		185		24		38		3
.69	.10695		.00677		.01973		.00075	
		195		23		39		2
.70	.10500		.00700		.01934		.00077	
		205		21		42		2
.71	−.10295		−.00721		+.01892		+.00079	
		215		18		43		2
.72	.10080		.00739		.01849		.00081	
		225		17		45		2
.73	.09855		.00756		.01804		.00083	
		235		14		46		1
.74	.09620		.00770		.01758		.00084	
		+245		—11		—49		+1
0.75	−.09375		−.00781		+.01709		+.00085	

Interval	Bessel's Coefficients for							
	J''		J'''		J^{iv}		J^{v}	
n	$\frac{n(n-1)}{2}$	Diff.	$\frac{n(n-1)(n-\frac{1}{2})}{6}$	Diff.	$\frac{(n+1)\ldots(n-2)}{24}$	Diff.	$\frac{(n+1)\ldots(n-\frac{1}{2})}{120}$	Diff.
0.75	−.09375		−.00781		+.01709		+.00085	
		+255		— 9		—50		+1
.76	.09120		.00790		.01659		.00086	
		265		7		52		+1
.77	.08855		.00797		.01607		.00087	
		275		4		54		0
.78	.08580		.00801		.01553		.00087	
		285		— 1		56		0
.79	.08295		.00802		.01497		.00087	
		295		+ 2		57		—1
.80	.08000		.00800		.01440		.00086	
		305		5		59		0
.81	−.07695		−.00795		+.01381		+.00086	
		315		8		60		—1
.82	.07380		.00787		.01321		.00085	
		325		11		62		2
.83	.07055		.00776		.01259		.00083	
		335		14		64		2
.84	.06720		.00762		.01195		.00081	
		345		18		65		2
.85	.06375		.00744		.01130		.00079	
		355		22		66		2
.86	−.06020		−.00722		+.01064		+.00077	
		365		25		68		3
.87	.05655		.00697		.00996		.00074	
		375		28		70		4
.88	.05280		.00669		.00926		.00070	
		385		33		70		3
.89	.04895		.00636		.00856		.00067	
		395		36		72		4
.90	.04500		.00600		.00784		.00063	
		405		40		74		5
.91	−.04095		−.00560		+.00710		+.00058	
		415		45		74		5
.92	.03680		.00515		.00636		.00053	
		425		48		76		5
.93	.03255		.00467		.00560		.00048	
		435		53		77		5
.94	.02820		.00414		.00483		.00043	
		445		58		78		7
.95	.02375		.00356		.00405		.00036	
		455		62		79		6
.96	−.01920		−.00294		+.00326		+.00030	
		465		66		80		7
.97	.01455		.00228		.00246		.00023	
		475		71		81		7
.98	.00980		.00157		.00165		.00016	
		485		76		82		8
.99	−.00495		−.00081		+.00083		+.00008	
		+495		+81		—83		—8
1.00	.00000		.00000		.00000		.00000	

Interval	Coefficients for						
	J''	J'''		J^{iv}		J^{v}	
n	$n-\frac{1}{2}$	$\frac{n^2}{2}-n+\frac{1}{3}$	Diff.	$\frac{n^3}{6}-\frac{3}{4}n^2+\frac{11}{12}n-\frac{1}{4}$	Diff.	$\frac{n^4}{24}-\frac{n^3}{3}+\dots+\frac{1}{5}$	Diff.
0.00	−0.50	+.33333		−.25000		+.20000	
			−995		+909		−825
.01	.49	.32338		.24091		.19175	
			985		894		807
.02	.48	.31353		.23197		.18368	
			975		880		790
.03	.47	.30378		.22317		.17578	
			965		865		773
.04	.46	.29413		.21452		.16805	
			955		850		757
.05	.45	.28458		.20602		.16048	
			945		836		740
.06	.44	+.27513		−.19766		+.15308	
			935		821		724
.07	.43	.26578		.18945		.14584	
			925		807		708
.08	.42	.25653		.18138		.13876	
			915		793		691
.09	.41	.24738		.17345		.13185	
			905		778		676
.10	.40	.23833		.16567		.12509	
			895		765		661
.11	.39	+.22938		−.15802		+.11848	
			885		751		645
.12	.38	.22053		.15051		.11203	
			875		737		630
.13	.37	.21178		.14314		.10573	
			865		723		615
.14	.36	.20313		.13591		.09958	
			855		710		600
.15	.35	.19458		.12881		.09358	
			845		696		585
.16	.34	+.18613		−.12185		+.08773	
			835		683		571
.17	.33	.17778		.11502		.08202	
			825		669		557
.18	.32	.16953		.10833		.07645	
			815		656		543
.19	.31	.16138		.10177		.07102	
			805		644		529
.20	.30	.15333		.09533		.06573	
			795		630		515
.21	.29	+.14538		−.08903		+.06058	
			785		617		502
.22	.28	.13753		.08286		.05556	
			775		605		488
.23	.27	.12978		.07681		.05068	
			765		591		475
.24	.26	.12213		.07090		.04593	
			−755		+580		−462
0.25	−0.25	+.11458		−.06510		+.04131	

Interval	Coefficients for						
	J''	J'''		J^{iv}		J^{v}	
n	$n-\frac{1}{2}$	$\frac{n^2}{2}-n+\frac{1}{3}$	Diff.	$\frac{n^3}{6}-\frac{3}{4}n^2+\frac{11}{12}n-\frac{1}{4}$	Diff.	$\frac{n^4}{24}-\frac{n^3}{3}+\dots+\frac{1}{5}$	Diff.
0.25	−0.25	+.11458		−.06510		+.04131	
			−745		+566		−449
.26	.24	.10713		.05944		.03682	
			735		555		437
.27	.23	.09978		.05389		.03245	
			725		542		424
.28	.22	.09253		.04847		.02821	
			715		529		412
.29	.21	.08538		.04318		.02409	
			705		518		400
.30	.20	.07833		.03800		.02009	
			695		506		388
.31	.19	+.07138		−.03294		+.01621	
			685		493		376
.32	.18	.06453		.02801		.01245	
			675		482		365
.33	.17	.05778		.02319		.00880	
			665		471		353
.34	.16	.05113		.01848		.00527	
			655		458		342
.35	.15	.04458		.01390		+.00185	
			645		448		330
.36	.14	+.03813		−.00942		−.00145	
			635		435		320
.37	.13	.03178		.00507		.00465	
			625		425		300
.38	.12	.02553		−.00082		.00774	
			615		413		298
.39	.11	.01938		+.00331		.01072	
			605		402		288
.40	.10	.01333		.00733		.01360	
			595		392		278
.41	.09	+.00738		+.01125		−.01638	
			585		380		267
.42	.08	+.00153		.01505		.01905	
			575		369		257
.43	.07	−.00422		.01874		.02162	
			565		359		248
.44	.06	.00987		.02233		.02410	
			555		348		238
.45	.05	.01542		.02581		.02648	
			545		338		228
.46	.04	−.02087		+.02919		−.02876	
			535		327		210
.47	.03	.02622		.03246		.03095	
			525		317		210
.48	.02	.03147		.03563		.03305	
			515		307		201
.49	−0.01	.03662		.03870		.03506	
			−505		+297		−192
0.50	0.00	−.04167		+.04167		−.03698	

Interval	Coefficients for						
	Δ''	Δ'''		Δ^{iv}		Δ^{v}	
n	$n-\frac{1}{2}$	$\frac{n^2}{2}-n+\frac{1}{3}$	Diff.	$\frac{n^3}{6}-\frac{3}{4}n^2+1\frac{1}{12}n-\frac{1}{4}$	Diff.	$\frac{n^4}{24}-\frac{n^3}{3}+\ldots+\frac{1}{5}$	Diff.
0.50	0.00	−.04167	−495	+.04167	+286	−.03698	−183
.51	+0.01	.04662	485	.04453	277	.03881	175
.52	.02	.05147	475	.04730	267	.04056	166
.53	.03	.05622	465	.04997	257	.04222	158
.54	.04	.06087	455	.05254	248	.04380	149
.55	.05	.06542	445	.05502	238	.04529	142
.56	.06	−.06987	435	+.05740	229	−.04671	134
.57	.07	.07422	425	.05969	220	.04805	126
.58	.08	.07847	415	.06189	210	.04931	118
.59	.09	.08262	405	.06399	201	.05049	111
.60	.10	.08667	395	.06600	192	.05160	104
.61	.11	−.09062	385	+.06792	183	−.05264	96
.62	.12	.09447	375	.06975	175	.05360	90
.63	.13	.09822	365	.07150	166	.05450	82
.64	.14	.10187	355	.07316	157	.05532	76
.65	.15	.10542	345	.07473	149	.05608	70
.66	.16	−.10887	335	+.07622	140	−.05678	62
.67	.17	.11222	325	.07762	132	.05740	57
.68	.18	.11547	315	.07894	124	.05797	50
.69	.19	.11862	305	.08018	115	.05847	44
.70	.20	.12167	295	.08133	108	.05891	38
.71	.21	−.12462	285	+.08241	100	−.05929	33
.72	.22	.12747	275	.08341	92	.05962	27
.73	.23	.13022	265	.08433	84	.05989	21
.74	.24	.13287	−255	.08517	+ 77	.06010	− 15
0.75	+0.25	−.13542		+.08594		−.06025	

Interval	Coefficients for						
	Δ''	Δ'''		Δ^{iv}		Δ^{v}	
n	$n-\frac{1}{2}$	$\frac{n^2}{2}-n+\frac{1}{3}$	Diff.	$\frac{n^3}{6}-\frac{3}{4}n^2+1\frac{1}{12}n-\frac{1}{4}$	Diff.	$\frac{n^4}{24}-\frac{n^3}{3}+\ldots+\frac{1}{5}$	Diff.
0.75	+0.25	−.13542	−245	+.08594	+69	−.06025	−11
.76	.26	.13787	235	.08663	62	.06036	− 5
.77	.27	.14022	225	.08725	54	.06041	0
.78	.28	.14247	215	.08779	47	.06041	+ 5
.79	.29	.14462	205	.08826	41	.06036	9
.80	.30	.14667	195	.08867	33	.06027	15
.81	.31	−.14862	185	+.08900	26	−.06012	19
.82	.32	.15047	175	.08926	20	.05993	23
.83	.33	.15222	165	.08946	12	.05970	28
.84	.34	.15387	155	.08958	+ 7	.05942	32
.85	.35	.15542	145	.08965	− 1	.05910	36
.86	.36	−.15687	135	+.08964	6	−.05874	40
.87	.37	.15822	125	.08958	13	.05834	44
.88	.38	.15947	115	.08945	20	.05790	47
.89	.39	.16062	105	.08925	25	.05743	52
.90	.40	.16167	95	.08900	31	.05691	55
.91	.41	−.16262	85	+.08869	38	−.05636	58
.92	.42	.16347	75	.08831	43	.05578	62
.93	.43	.16422	65	.08788	48	.05516	65
.94	.44	.16487	55	.08740	55	.05451	68
.95	.45	.16542	45	.08685	59	.05383	71
.96	.46	−.16587	35	+.08626	66	−.05312	74
.97	.47	.16622	25	.08560	70	.05238	76
.98	.48	.16647	15	.08490	76	.05162	80
.99	.49	.16662	− 5	.08414	−81	.05082	+82
1.00	+0.50	−.16667		+.08333		−.05000	

NOTE: — The coefficient for A'' = the given argument n.

Interval	Coefficients for						Interval	Coefficients for					
	J'''		J^{iv}		J^{v}			J'''		J^{iv}		J^{v}	
n	$\frac{n^2}{2}-\frac{1}{6}$	Diff.	$\frac{n^3}{6}-\frac{n}{12}$	Diff.	$\frac{n^4}{24}-\frac{n^2}{8}+\frac{1}{30}$	Diff.	n	$\frac{n^2}{2}-\frac{1}{6}$	Diff.	$\frac{n^3}{6}-\frac{n}{12}$	Diff.	$\frac{n^4}{24}-\frac{n^2}{8}+\frac{1}{30}$	Diff.
0.00	−.16667	+ 5	.00000	−83	+.03333	− 1	0.25	−.13542	+255	−.01823	−51	+.02568	− 61
.01	.16662	15	−.00083	84	.03332	4	.26	.13287	265	.01874	48	.02507	63
.02	.16647	25	.00167	83	.03328	6	.27	.13022	275	.01922	45	.02444	65
.03	.16622	35	.00250	82	.03322	9	.28	.12747	285	.01967	43	.02379	67
.04	.16587	45	.00332	83	.03313	11	.29	.12462	295	.02010	40	.02312	70
.05	.16542	55	.00415	81	.03302	14	.30	.12167	305	.02050	37	.02242	71
.06	−.16487	65	−.00496	82	+.03288	16	.31	−.11862	315	−.02087	34	+.02171	74
.07	.16422	75	.00578	80	.03272	18	.32	.11547	325	.02121	30	.02097	76
.08	.16347	85	.00658	80	.03254	22	.33	.11222	335	.02151	27	.02021	77
.09	.16262	95	.00738	79	.03232	23	.34	.10887	345	.02178	24	.01944	79
.10	.16167	105	.00817	77	.03209	26	.35	.10542	355	.02202	20	.01865	82
.11	−.16062	115	−.00894	77	+.03183	29	.36	−.10187	365	−.02222	17	+.01783	83
.12	.15947	125	.00971	76	.03154	31	.37	.09822	375	.02239	13	.01700	85
.13	.15822	135	.01047	74	.03123	33	.38	.09447	385	.02252	9	.01615	87
.14	.15687	145	.01121	73	.03090	36	.39	.09062	395	.02261	6	.01528	88
.15	.15542	155	.01194	71	.03054	38	.40	.08667	405	.02267	− 1	.01440	90
.16	−.15387	165	−.01265	70	+.03016	40	.41	−.08262	415	−.02268	+ 3	+.01350	92
.17	.15222	175	.01335	68	.02976	43	.42	.07847	425	.02265	7	.01258	93
.18	.15047	185	.01403	66	.02933	45	.43	.07422	435	.02258	11	.01165	95
.19	.14862	195	.01469	64	.02888	48	.44	.06987	445	.02247	16	.01070	97
.20	.14667	205	.01533	63	.02840	50	.45	.06542	455	.02231	20	.00973	98
.21	−.14462	215	−.01596	60	+.02790	52	.46	−.06087	465	−.02211	25	+.00875	100
.22	.14247	225	.01656	58	.02738	54	.47	.05622	475	.02186	29	.00775	100
.23	.14022	235	.01714	56	.02684	57	.48	.05147	485	.02157	34	.00675	103
.24	.13787	+245	.01770	−53	.02627	−59	.49	.04662	+495	.02123	+40	.00572	−103
0.25	−.13542		−.01823		+.02568		0.50	−.04167		−.02083		+.00469	

Interval	Coefficients for					
	Δ'''		Δ^{iv}		Δ^{v}	
n	$\frac{n^2}{2}-\frac{1}{6}$	Diff.	$\frac{n^3}{6}-\frac{n}{12}$	Diff.	$\frac{n^4}{24}-\frac{n^2}{8}+\frac{1}{30}$	Diff.
0.50	−.04167	+505	−.02083	+ 44	+.00469	−105
.51	.03662	515	.02039	49	.00364	106
.52	.03147	525	.01990	55	.00258	107
.53	.02622	535	.01935	59	.00151	108
.54	.02087	545	.01876	66	+.00043	110
.55	.01542	555	.01810	70	−.00067	110
.56	−.00987	565	−.01740	77	−.00177	111
.57	−.00422	575	.01663	82	.00288	112
.58	+.00153	585	.01581	87	.00400	113
.59	.00738	595	.01494	94	.00513	114
.60	.01333	605	.01400	100	.00627	114
.61	+.01938	615	−.01300	105	−.00741	115
.62	.02553	625	.01195	112	.00856	116
.63	.03178	635	.01083	119	.00972	116
.64	.03813	645	.00964	124	.01088	116
.65	.04458	655	.00840	132	.01204	117
.66	+.05113	665	−.00708	137	−.01321	117
.67	.05778	675	.00571	145	.01438	118
.68	.06453	685	.00426	151	.01556	117
.69	.07138	695	.00275	158	.01673	118
.70	.07833	705	−.00117	166	.01791	118
.71	+.08538	715	+.00049	172	−.01909	118
.72	.09253	725	.00221	179	.02027	118
.73	.09978	735	.00400	187	.02145	117
.74	.10713	+745	.00587	+194	.02262	−118
0.75	+.11458		+.00781		−.02380	

Interval	Coefficients for					
	Δ'''		Δ^{iv}		Δ^{v}	
n	$\frac{n^2}{2}-\frac{1}{6}$	Diff.	$\frac{n^3}{6}-\frac{n}{12}$	Diff.	$\frac{n^4}{24}-\frac{n^2}{8}+\frac{1}{30}$	Diff.
0.75	+.11458	+755	+.00781	+202	−.02380	−117
.76	.12213	765	.00983	209	.02497	116
.77	.12978	775	.01192	217	.02613	116
.78	.13753	785	.01409	225	.02729	116
.79	.14538	795	.01634	233	.02845	115
.80	.15333	805	.01867	240	.02960	114
.81	+.16138	815	+.02107	249	−.03074	114
.82	.16953	825	.02356	257	.03188	112
.83	.17778	835	.02613	265	.03300	112
.84	.18613	845	.02878	274	.03412	111
.85	.19458	855	.03152	282	.03523	109
.86	+.20313	865	+.03434	291	−.03632	109
.87	.21178	875	.03725	300	.03741	107
.88	.22053	885	.04025	308	.03848	106
.89	.22938	895	.04333	317	.03954	104
.90	.23833	905	.04650	326	.04058	103
.91	+.24738	915	+.04976	335	−.04161	101
.92	.25653	925	.05311	345	.04262	99
.93	.26578	935	.05656	354	.04361	98
.94	.27513	945	.06010	363	.04459	95
.95	.28458	955	.06373	373	.04554	94
.96	+.29413	965	+.06746	382	−.04648	91
.97	.30378	975	.07128	392	.04739	89
.98	.31353	985	.07520	402	.04828	87
.99	.32338	+995	.07922	+411	.04915	− 85
1.00	+.33333		+.08333		−.05000	

NOTE: — The coefficient for Δ'' = the given argument n.

Interval	Coefficients for						
	Δ''	Δ'''		Δ^{iv}		Δ^{v}	
n	$n-\frac{1}{2}$	$\frac{n^2}{2}-\frac{n}{2}+\frac{1}{12}$	Diff.	$\frac{n^3}{6}-\frac{n^2}{4}-\frac{n}{12}+\frac{1}{12}$	Diff.	$\frac{n^4}{24}-\frac{n^3}{12}+\frac{n}{24}-\frac{1}{120}$	Diff.
0.00	−0.50	+.08333	−495	+.08333	− 85	−.00833	+41
.01	.49	.07838	485	.08248	91	.00792	42
.02	.48	.07353	475	.08157	96	.00750	41
.03	.47	.06878	465	.08061	100	.00709	42
.04	.46	.06413	455	.07961	105	.00667	41
.05	.45	.05958	445	.07856	109	.00626	41
.06	.44	+.05513	435	+.07747	114	−.00585	41
.07	.43	.05078	425	.07633	118	.00544	40
.08	.42	.04653	415	.07515	122	.00504	40
.09	.41	.04238	405	.07393	126	.00464	39
.10	.40	.03833	395	.07267	131	.00425	40
.11	.39	+.03438	385	+.07136	134	−.00385	38
.12	.38	.03053	375	.07002	138	.00347	38
.13	.37	.02678	365	.06864	142	.00309	38
.14	.36	.02313	355	.06722	145	.00271	37
.15	.35	.01958	345	.06577	149	.00234	36
.16	.34	+.01613	335	+.06428	152	−.00198	36
.17	.33	.01278	325	.06276	155	.00162	34
.18	.32	.00953	315	.06121	159	.00128	35
.19	.31	.00638	305	.05962	162	.00093	33
.20	.30	.00333	295	.05800	165	.00060	33
.21	.29	+.00038	285	+.05635	168	−.00027	31
.22	.28	−.00247	275	.05467	170	+.00004	31
.23	.27	.00522	265	.05297	173	.00035	30
.24	.26	.00787	−255	.05124	−176	.00065	+29
0.25	−0.25	−.01042		+.04948		+.00094	

Interval	Coefficients for						
	Δ''	Δ'''		Δ^{iv}		Δ^{v}	
n	$n-\frac{1}{2}$	$\frac{n^2}{2}-\frac{n}{2}+\frac{1}{12}$	Diff.	$\frac{n^3}{6}-\frac{n^2}{4}-\frac{n}{12}+\frac{1}{12}$	Diff.	$\frac{n^4}{24}-\frac{n^3}{12}+\frac{n}{24}-\frac{1}{120}$	Diff.
0.25	−0.25	−.01042	−245	+.04948	−178	+.00094	+29
.26	.24	.01287	235	.04770	181	.00123	27
.27	.23	.01522	225	.04589	183	.00150	26
.28	.22	.01747	215	.04406	185	.00176	25
.29	.21	.01962	205	.04221	188	.00201	24
.30	.20	.02167	195	.04033	189	.00225	24
.31	.19	−.02362	185	+.03844	191	+.00249	22
.32	.18	.02547	175	.03653	193	.00271	21
.33	.17	.02722	165	.03460	195	.00292	19
.34	.16	.02887	155	.03265	196	.00311	19
.35	.15	.03042	145	.03069	198	.00330	18
.36	.14	−.03187	135	+.02871	199	+.00348	16
.37	.13	.03322	125	.02672	201	.00364	16
.38	.12	.03447	115	.02471	202	.00380	14
.39	.11	.03562	105	.02269	202	.00394	13
.40	.10	.03667	95	.02067	204	.00407	11
.41	.09	−.03762	85	+.01863	205	+.00418	11
.42	.08	.03847	75	.01658	205	.00429	9
.43	.07	.03922	65	.01453	207	.00438	8
.44	.06	.03987	55	.01246	206	.00446	7
.45	.05	.04042	45	.01040	208	.00453	6
.46	.04	−.04087	35	+.00832	207	+.00459	4
.47	.03	.04122	25	.00625	208	.00463	3
.48	.02	.04147	15	.00417	200	.00466	2
.49	−0.01	.04162	− 5	+.00208	−208	.00468	+ 1
0.50	0.00	−.04167		.00000		+.00469	

Interval	Coefficients for						
	Δ''	Δ'''		Δ^{iv}		Δ^{v}	
n	$n-\frac{1}{2}$	$\frac{n^2}{2}-\frac{n}{2}+\frac{1}{12}$	Diff.	$\frac{n^3}{6}-\frac{n^2}{4}-\frac{n}{12}+\frac{1}{12}$	Diff.	$\frac{n^4}{24}-\frac{n^3}{12}+\frac{n}{24}-\frac{1}{120}$	Diff.
0.50	0.00	−.04167	+5	.00000	−208	+.00469	−1
.51	+.01	.04162	15	−.00208	209	.00468	2
.52	.02	.04147	25	.00417	208	.00466	3
.53	.03	.04122	35	.00625	207	.00463	4
.54	.04	.04087	45	.00832	208	.00459	6
.55	.05	.04042	55	.01040	206	.00453	7
.56	.06	−.03987	65	−.01246	207	+.00446	8
.57	.07	.03922	75	.01453	205	.00438	9
.58	.08	.03847	85	.01658	205	.00429	11
.59	.09	.03762	95	.01863	204	.00418	11
.60	.10	.03667	105	.02067	202	.00407	13
.61	.11	−.03562	115	−.02269	202	+.00394	14
.62	.12	.03447	125	.02471	201	.00380	16
.63	.13	.03322	135	.02672	199	.00364	16
.64	.14	.03187	145	.02871	198	.00348	18
.65	.15	.03042	155	.03069	196	.00330	19
.66	.16	−.02887	165	−.03265	195	+.00311	19
.67	.17	.02722	175	.03460	193	.00292	21
.68	.18	.02547	185	.03653	191	.00271	22
.69	.19	.02362	195	.03844	189	.00249	24
.70	.20	.02167	205	.04033	188	.00225	24
.71	.21	−.01962	215	−.04221	185	+.00201	25
.72	.22	.01747	225	.04406	183	.00176	26
.73	.23	.01522	235	.04589	181	.00150	27
.74	.24	.01287	+245	.04770	−178	.00123	−29
0.75	+0.25	−.01042		−.04948		+.00094	

Interval	Coefficients for						
	Δ''	Δ'''		Δ^{iv}		Δ^{v}	
n	$n-\frac{1}{2}$	$\frac{n^2}{2}-\frac{n}{2}+\frac{1}{12}$	Diff.	$\frac{n^3}{6}-\frac{n^2}{4}-\frac{n}{12}+\frac{1}{12}$	Diff.	$\frac{n^4}{24}-\frac{n^3}{12}+\frac{n}{24}-\frac{1}{120}$	Diff.
0.75	+0.25	−.01042	+255	−.04948	−176	+.00094	−29
.76	.26	.00787	265	.05124	173	.00065	30
.77	.27	.00522	275	.05297	170	.00035	31
.78	.28	−.00247	285	.05467	168	+.00004	31
.79	.29	+.00038	295	.05635	165	−.00027	33
.80	.30	.00333	305	.05800	162	.00060	33
.81	.31	+.00638	315	−.05962	159	−.00093	35
.82	.32	.00953	325	.06121	155	.00128	34
.83	.33	.01278	335	.06276	152	.00162	36
.84	.34	.01613	345	.06428	149	.00198	36
.85	.35	.01958	355	.06577	145	.00234	37
.86	.36	+.02313	365	−.06722	142	−.00271	38
.87	.37	.02678	375	.06864	138	.00309	38
.88	.38	.03053	385	.07002	134	.00347	38
.89	.39	.03438	395	.07136	131	.00385	40
.90	.40	.03833	405	.07267	126	.00425	30
.91	.41	+.04238	415	−.07393	122	−.00464	40
.92	.42	.04653	425	.07515	118	.00504	40
.93	.43	.05078	435	.07633	114	.00544	41
.94	.44	.05513	445	.07747	109	.00585	41
.95	.45	.05958	455	.07856	105	.00626	41
.96	.46	+.06413	465	−.07961	100	−.00667	42
.97	.47	.06878	475	.08061	96	.00709	41
.98	.48	.07353	485	.08157	91	.00750	42
.99	.49	.07838	+495	.08248	−85	.00792	−41
1.00	+0.50	+.08333		−.08333		−.00833	

The tabular quantities are in units of the sixth decimal.

x	0.00	.01	.02	.03	.04	.05	.06	.07	.08	.09	.10	.11	.12	.13	.14	.15	.16	.17	.18	.19	0.20	x
	Values of Argument K.																					
0.000	0	0	0	0	0	0	0	0	0	0	0	0	0	0	0	0	0	0	0	0	0	0.000
.001	0	0	0	0	0	0	0	0	0	0	0	0	0	0	0	0	0	0	0	0	0	.001
.002	0	0	0	0	0	0	0	0	0	0	0	0	0	0	0	0	0	0	0	0	0	.002
.003	0	0	0	0	0	0	0	0	0	0	0	0	1	1	1	1	1	1	1	1	1	.003
.004	0	0	0	0	0	0	0	1	1	1	1	1	1	1	1	1	1	1	1	2	2	.004
.005	0	0	0	0	0	1	1	1	1	1	1	1	2	2	2	2	2	2	2	2	3	.005
.006	0	0	0	1	1	1	1	1	1	2	2	2	2	2	3	3	3	3	3	3	4	.006
.007	0	0	0	1	1	1	1	2	2	2	2	3	3	3	3	4	4	4	4	5	5	.007
.008	0	0	1	1	1	2	2	2	3	3	3	4	4	4	4	5	5	5	6	6	6	.008
.009	0	0	1	1	2	2	2	3	3	4	4	4	5	5	6	6	6	7	7	8	8	.009
.010	0	0	1	1	2	2	3	3	4	4	5	6	6	7	7	8	8	9	9	10	10	.010
.011	0	1	1	2	2	3	4	4	5	5	6	7	7	8	8	9	10	10	11	11	12	.011
.012	0	1	1	2	3	4	4	5	6	6	7	8	9	9	10	11	12	12	13	14	14	.012
.013	0	1	2	3	3	4	5	6	7	8	8	9	10	11	12	13	14	14	15	16	17	.013
.014	0	1	2	3	4	5	6	7	8	9	10	11	12	13	14	15	16	17	18	19	20	.014
.015	0	1	2	3	4	6	7	8	9	10	11	12	14	15	16	17	18	19	20	21	23	.015
.016	0	1	3	4	5	6	8	9	10	12	13	14	15	17	18	19	20	22	23	24	26	.016
.017	0	1	3	4	6	7	9	10	12	13	14	16	17	19	20	22	23	25	26	27	29	.017
.018	0	2	3	5	6	8	10	11	13	15	16	18	19	21	23	24	26	28	29	31	32	.018
.019	0	2	4	5	7	9	11	13	14	16	18	20	22	23	25	27	29	31	32	34	36	.019
.020	0	2	4	6	8	10	12	14	16	18	20	22	24	26	28	30	32	34	36	38	40	.020
.021	0	2	4	7	9	11	13	15	18	20	22	24	26	29	31	33	35	37	40	42	44	.021
.022	0	2	5	7	10	12	15	17	19	22	24	27	29	31	34	36	39	41	44	46	48	.022
.023	0	3	5	8	11	13	16	19	21	24	26	29	32	34	37	40	42	45	48	50	53	.023
.024	0	3	6	9	12	14	17	20	23	26	29	32	35	37	40	43	46	49	52	55	58	.024
.025	0	3	6	9	13	16	19	22	25	28	31	34	38	41	44	47	50	53	56	59	63	0.025
	0.00	.01	.02	.03	.04	.05	.06	.07	.08	.09	.10	.11	.12	.13	.14	.15	.16	.17	.18	.19	0.20	

GIVING y: TO BE USED IN FINDING n WHEN F_n IS GIVEN.

NOTE.—The quantity y has the same sign as argument K.

x	Values of Argument K. 0.00	.01	.02	.03	.04	.05	.06	.07	.08	.09	.10	.11	.12	.13	.14	.15	.16	.17	.18	.19	0.20	x
0.025	0	3	6	9	13	16	19	22	25	28	31	34	38	41	44	47	50	53	56	59	63	0.025
.026	0	3	7	10	14	17	20	24	27	30	34	37	41	44	47	51	54	57	61	64	68	.026
.027	0	4	7	11	15	18	22	26	29	33	36	40	44	47	51	55	58	62	66	69	73	.027
.028	0	4	8	12	16	20	24	27	31	35	39	43	47	51	55	59	63	67	71	74	78	.028
.029	0	4	8	13	17	21	25	29	34	38	42	46	50	55	59	63	67	71	76	80	84	.029
.030	0	5	9	14	18	23	27	32	36	41	45	50	54	58	63	68	72	76	81	86	90	.030
.031	0	5	10	14	19	24	29	34	38	43	48	53	58	62	67	72	77	82	86	91	96	.031
.032	0	5	10	15	20	26	31	36	41	46	51	56	61	67	72	77	82	87	92	97	102	.032
.033	0	5	11	16	22	27	33	38	44	49	54	60	65	71	76	82	87	93	98	103	109	.033
.034	0	6	12	17	23	29	35	40	46	52	58	64	69	75	81	87	92	98	104	110	116	.034
.035	0	6	12	18	25	31	37	43	49	55	61	67	74	80	86	92	98	104	110	116	123	.035
.036	0	6	13	19	26	32	39	45	52	58	65	71	78	84	91	97	104	110	117	123	130	.036
.037	0	7	14	21	27	34	41	48	55	62	68	75	82	89	96	103	110	116	123	130	137	.037
.038	0	7	14	22	29	36	43	51	58	65	72	79	87	94	101	108	116	123	130	137	144	.038
.039	0	8	15	23	30	38	46	53	61	68	76	84	91	99	106	114	122	129	137	144	152	.039
.040	0	8	16	24	32	40	48	56	64	72	80	88	96	104	112	120	128	136	144	152	160	.040
.041	0	8	17	25	34	42	50	59	67	76	84	92	101	109	118	126	134	143	151	160	168	.041
.042	0	9	18	26	35	44	53	62	71	79	88	97	106	115	123	132	141	150	159	168	176	.042
.043	0	9	18	28	37	46	55	65	74	83	92	102	111	120	129	139	148	157	166	176	185	.043
.044	0	10	19	29	39	48	58	68	77	87	97	106	116	126	136	145	155	165	174	184	194	.044
.045	0	10	20	30	40	51	61	71	81	91	101	111	122	132	142	152	162	172	182	192	203	.045
.046	0	11	21	32	42	53	63	74	85	95	106	116	127	138	148	159	169	180	190	201	212	.046
.047	0	11	22	33	44	55	66	77	88	99	110	121	133	144	155	166	177	188	199	210	221	.047
.048	0	12	23	35	46	58	69	81	92	104	115	127	138	150	161	173	184	196	207	219	230	.048
.049	0	12	24	36	48	60	72	84	96	108	120	132	144	156	168	180	192	204	216	228	240	.049
0.050	0	12	25	38	50	62	75	88	100	112	125	138	150	162	175	188	200	212	225	238	250	0.050
	0.00	.01	.02	.03	.04	.05	.06	.07	.08	.09	.10	.11	.12	.13	.14	.15	.16	.17	.18	.19	0.20	

The tabular quantities are in units of the sixth decimal.

GIVING y: TO BE USED IN FINDING n WHEN F_n IS GIVEN.

NOTE. — The quantity y has the same sign as argument K.

COEFFICIENTS FOR COMPUTING

$$F_n = F_0 + n\omega\left(F_0' + \frac{n}{2}\alpha + \mathrm{B}\beta_0 + \Gamma\gamma\right).$$

n	$\mathrm{B} \equiv \frac{n^2}{6}$	Diff.	$\Gamma \equiv \frac{n}{12}\left(\frac{n^2}{2}-1\right)$	Diff.
0.00	0.0000		0.0000	
		+1		−17
.02	+ .0001		− .0017	
		2		16
.04	.0003		.0033	
		3		17
.06	.0006		.0050	
		5		16
.08	.0011		.0066	
		6		17
.10	.0017		.0083	
		7		16
.12	+ .0024		− .0099	
		9		17
.14	.0033		.0116	
		10		16
.16	.0043		.0132	
		11		16
.18	.0054		.0148	
		13		15
.20	.0067		.0163	
		14		16
.22	+ .0081		− .0179	
		15		15
.24	.0096		.0194	
		17		15
.26	.0113		.0209	
		18		15
.28	.0131		.0224	
		19		15
.30	.0150		.0239	
		21		14
.32	+ .0171		− .0253	
		22		14
.34	.0193		.0267	
		23		14
.36	.0216		.0281	
		25		13
.38	.0241		.0294	
		26		13
.40	.0267		.0307	
		27		12
.42	+ .0294		− .0319	
		29		12
.44	.0323		.0331	
		30		12
.46	.0353		.0343	
		31		11
.48	.0384		.0354	
		33		11
.50	.0417		.0365	
		34		10
.52	+ .0451		− .0375	
		35		9
.54	.0486		.0384	
		37		9
.56	.0523		.0393	
		38		9
.58	.0561		.0402	
		+39		− 8
0.60	+0.0600		−0.0410	

BIBLIOGRAPHY.

List of the Principal Papers, Memoires, etc., upon the Subjects of Interpolation and Mechanical Quadrature.

Åstrand (J. J.). Vierteljahrsschrift der Astronomischen Gesellschaft, Vol. X (1875), p. 279.

Baillaud (B.). Annales de l'Observatoire de Toulouse, Vol. II, p. B. 1.

Bienaymé (Jules). Liouville, Journal de Mathématiques, Vol. XVIII (1853), p. 299.

Boole (George). A Treatise on the Calculus of Finite Differences, Chapters II and III.

Brassinne (E.). Liouville, Journal de Mathématiques, Vol. XI (1846), p. 177.

Brünnow (F.). Lehrbuch der Sphärischen Astronomie, p. 18.

Cauchy (Augustin). (*a*) Liouville, Journal de Mathématiques, Vol. II (1837), p. 193.
(*b*) Comptes Rendus, Vol. XI (1840), p. 775.
(*c*) Ibid., Vol. XIX (1844), p. 1183.
(*d*) Connaissance des Temps, 1852 (Additions), p. 129.

Chauvenet (Wm.). Spherical and Practical Astronomy, Vol. I, p. 79.

Davis (C. H.). The Mathematical Monthly (Cambridge, Mass.), Vol. II (1860), p. 276.

Doolittle (C. L.). A Treatise on Practical Astronomy, p. 70.

Encke (J. F.). (*a*) Berliner Astronomisches Jahrbuch, 1830, p. 265.
(*b*) Ibid., 1837, p. 251.
(*c*) Ibid., 1852, p. 330.
(*d*) Ibid., 1862, p. 313.

Ferrel (William). The Mathematical Monthly (Cambridge, Mass.), Vol. III (1861), p. 377.

Gauss (Carl F.). Werke, Vol. III, p. 265.

Grunert (J. A.). (*a*) Archiv der Mathematik und Physik, Vol. XIV (1850), p. 225.
(*b*) Ibid., Vol. XX (1853), p. 361.

Hansen (P. A.). (*a*) Abhandlungen der Königlich Sächsischen Gesellschaft der Wissenschaften (Leipzig), Vol. XI (1865), p. 505.
(*b*) Tables de la Lune, p. 68.

Herr *und* **Tinter.** Lehrbuch der Sphärischen Astronomie, Chapter II.

Jacobi (C. G. J.). (*a*) Crelle's Journal, Vol. I (1826), p. 301.
(*b*) Ibid., Vol. XXX (1846), p. 127.

Klinkerfues (W.). Theoretische Astronomie (2d edition, 1899), pp. 67 and 490.

Klügel's Mathematisches Wörterbuch. See article "Einschalten," which includes a brief history of the subject.

Lagrange (J. L.). (*a*) Œuvres, Vol. V, p. 663.
(*b*) Ibid., Vol. VII, p. 535.

Laplace (P. S.). Mécanique Céleste, Vol. IV, pp. 204–207.

Le Verrier (U. J.). Annales de l'Observatoire de Paris (Mémoires), Vol. I, pp. 121, 129, 151, 154.

Loomis (Elias). Practical Astronomy, p. 202.

Maurice (Fréd.). Connaissance des Temps, 1847 (Additions), p. 181.

Merrifield (C. W.). British Association Report, Vol. L (1880), p. 321.

Newcomb (Simon). Logarithmic and Other Mathematical Tables, p. 56.

Newton (Isaac). Principia, Book III, Lemma V.

Olivier (Louis). Crelle's Journal, Vol. II (1827), p. 252.

Oppolzer (T. R.). Lehrbuch zur Bahnbestimmung, Vol. II, pp. 1 and 596.

Radau (Rodolphe). (*a*) Liouville, Journal de Mathématiques, 3d Series, Vol. VI (1880), p. 283.
(*b*). Bulletin Astronomique, Vol. VIII (1891), pp. 273, 325, 376, 425.

Rees's Cyclopedia. See article "Interpolation."

Sawitsch (A.). Abriss der Practischen Astronomie, Vol. II, p. 416; or see the one volume edition, p. 818.

Tisserand (F.). (*a*) Comptes Rendus, Vol. LXVIII (1869), p. 1101.
(*b*) Ibid., Vol. LXX (1870), p. 678.
(*c*) Traité de Mécanique Céleste, Vol. IV, Chapters X and XI.

Valentiner (W.). Handwörterbuch der Astronomie, Vol. II, pp. 41 and 618.

Watson (James C.). Theoretical Astronomy, pp. 112, 335, 435.

Weddle (Thomas). Cambridge and Dublin Mathematical Journal, Vol. IX (1854), p. 79.

www.ingramcontent.com/pod-product-compliance
Lightning Source LLC
Chambersburg PA
CBHW021527210326
41599CB00012B/1404
* 9 7 8 3 7 4 3 3 3 5 3 9 4 *